Hands-On Internet of Things with MQTT

Build connected IoT devices with Arduino and MQ Telemetry Transport (MQTT)

Tim Pulver

BIRMINGHAM - MUMBAI

Hands-On Internet of Things with MQTT

Commissioning Editor: Gebin George
Acquisition Editor: Rohit Rajkumar
Content Development Editor: Ronn Kurien
Senior Editor: Rahul Dsouza
Technical Editor: Mohd Riyan Khan
Copy Editor: Safis Editing
Project Coordinator: Vaidehi Sawant
Proofreader: Safis Editing
Indexer: Pratik Shirodkar
Production Designer: Jyoti Chauhan

First published: October 2019

Production reference: 1041019

Published by Packt Publishing Ltd.
Livery Place
35 Livery Street
Birmingham
B3 2PB, UK.

ISBN 978-1-78934-178-2

www.packt.com

Packt.com

Subscribe to our online digital library for full access to over 7,000 books and videos, as well as industry leading tools to help you plan your personal development and advance your career. For more information, please visit our website.

Why subscribe?

- Spend less time learning and more time coding with practical eBooks and Videos from over 4,000 industry professionals

- Improve your learning with Skill Plans built especially for you

- Get a free eBook or video every month

- Fully searchable for easy access to vital information

- Copy and paste, print, and bookmark content

Did you know that Packt offers eBook versions of every book published, with PDF and ePub files available? You can upgrade to the eBook version at www.packt.com and as a print book customer, you are entitled to a discount on the eBook copy. Get in touch with us at customercare@packtpub.com for more details.

At www.packt.com, you can also read a collection of free technical articles, sign up for a range of free newsletters, and receive exclusive discounts and offers on Packt books and eBooks.

Contributors

About the author

Tim Pulver is a Berlin-based freelance interaction designer and developer. In his work, he combines his physical prototyping experience and knowledge of interface design with modern technologies such as 3D printing, laser cutting, web technologies, and machine learning to create unique interactive experiences.

In recent years, he has worked on interactive data visualizations, web-based audiovisual experiences, musical interfaces, and cables— an innovative browser-based visual programming language that enables the creation of interactive audiovisual prototypes without writing any code.

He holds a Bachelor of Arts degree in interface design from the University of Applied Sciences in Potsdam, Germany.

I wish to thank Naree. Without your support, this book would not have been possible.

About the reviewer

Federico Gonzalez is an Argentina-based cooperative developer and teacher. He studied information systems engineering at UTN with a focus on software development. He works at Devecoop (his cooperative). He has worked on many projects, with his current focus on developing software and teaching React.js. He has contributed to open source projects such as Lelylan (an IoT cloud platform with the microservices architecture) and eventoL (conference and install fest management software), and has made some minor contributions to projects with Docker environment configuration, Python, and JavaScript code. He has also presented various workshops at universities, conferences, and companies in Argentina featuring React.js, Python, Docker, open source, free software, and cooperatives.

Packt is searching for authors like you

If you're interested in becoming an author for Packt, please visit `authors.packtpub.com` and apply today. We have worked with thousands of developers and tech professionals, just like you, to help them share their insight with the global tech community. You can make a general application, apply for a specific hot topic that we are recruiting an author for, or submit your own idea.

Table of Contents

Preface

First and foremost, I would like to thank you for purchasing this book and going on a journey of **Internet of Things (IoT)** prototyping with me.

In my teenage years, before the first Arduino was invented, I had a friend who was knowledgeable about electronics. He was considered a nerd and really knew what he was doing. Looking at his soldered prototypes, which consisted of a lot of cables, chips, and other electronic components that I didn't know about at the time, I was fascinated. I did not have any idea where to start building electronic prototypes myself; it seemed like such a huge field that there was no way for me to find an entry point. At that time, it was very far from my imagination that I would later be able to create prototypes myself, and, even more so, that I would write a book about the topic one day.

A lot of things have changed since then. One of those things is the creation of Arduino and Processing, which made it possible to learn programming and hardware prototyping without any prior knowledge. The field opened up to designers, artists, and makers. It is now easier than ever to get started with hardware prototyping and bring your own ideas into reality.

While building physical projects alone creates a lot of possibilities, being able to use internet-connected devices and connect them to each other and the rest of the web opens up a whole new world of possibilities.

There are many ways to connect two internet-connected devices and exchange information between them. One of these possibilities, which in my opinion is the easiest and most open, is MQTT. Being *open* in this sense means that third-party developers can create apps and libraries for it. There is a vast ecosystem of tools, libraries, and apps that all speak the same language of MQTT.

Building internet-connected devices has never been easier, and using MQTT to build them makes it possible to prototype even faster, while also having access to all these third-party tools.

When I was initially looking into possibilities for how to let my prototypes talk to each other over the internet, MQTT stood out, but I could not find much information about it. It seemed like a niche topic and I could not understand why.

If you want to create your own inventions, experimental prototypes, and custom devices, getting to know MQTT is a great decision, and I hope that, by reading this book, you will benefit from a smooth entry into the world of IoT prototyping using MQTT.

Welcome on this journey!

Who this book is for

This book is ideal for readers who want to start creating internet-connected projects themselves, but only have a little bit of experience of programming and using the Arduino IDE. You do not need to have a computer science background to follow along. Simply curiosity and a basic understanding of programming in Arduino will be enough. If you understand basic programming concepts such as variables, loops, conditions, and functions, then you will have no problem following along.

Some of the chapters touch on a lot of different topics. I do not expect you to understand every single bit in detail. The goal here is to give you a general overview of related technologies and techniques, so you can dive into any of the topics after finishing the book. You can look at it as a starting point for IoT prototyping.

For an intermediate or professional programmer, the pace of the hands-on projects in this book might feel a bit slow, but there is plenty of relevant information for you in the theoretical chapters as well, including how to use MQTT.

What this book covers

Chapter 1, *The Internet of Things in a Nutshell*, gives you an overview of fields where IoT plays an important role. It introduces you to smart cars, the **Industrial Internet of Things (IIoT)**, and smart homes.

Chapter 2, *Basic Architecture of an IoT Prototype*, introduces you to related technologies and concepts, as well as microcontroller recommendations for IoT prototyping. This chapter touches on a lot of topics and should be seen as a starting point for your own learning, after which you can dig deeper into the topics that interest you.

Chapter 3, *Getting Started with MQTT*, explains the concepts behind MQTT, the lightweight IoT protocol we will be using throughout the entire book. While we only use a subset of the features explained in this chapter during the hands-on projects, you can use it as a reference to go back to if you want to create more advanced MQTT projects later. Feel free to create a bookmark for this chapter, as you will hopefully re-read fragments from it.

Chapter 4, *Setting Up a Lab Environment*, explains how to install the software needed for the hands-on projects. It also includes a shopping list with all the necessary hardware for the projects.

Chapter 5, *Building Your Own Automatic Pet Food Dispenser*, is the first hands-on project in the book. It will show you, using a servo motor, how to control an automatic dispenser using MQTT, either from your computer or smartphone using one of the apps introduced in Chapter 3, *Getting Started with MQTT*. You can fill the dispenser with pet food, sweets, cereals, or whatever you like.

Chapter 6, *Building a Smart E-Ink To-Do List*, is the second hands-on project in the book. In this chapter, you will learn how to use an energy-efficient e-ink display and send text to it using MQTT. The display could, for example, be hung next to your front door to remind you to take out the trash or buy milk.

Chapter 7, *Building a Smart Productivity Cube, Part 1*, is the third and final hands-on project in the book. It explains how to create an orientation sensor from scratch using simple electrical components called tilt switches. The cube can then be used to record the time you spend on various activities. Because all three projects use MQTT, they can also be chained together, such that activating the smart productivity cube might show some text on the e-ink display and activate the food dispenser.

Chapter 8, *Building a Smart Productivity Cube, Part 2*, is the second part of *building a smart productivity cube*. Based on the prototype that we built together in the previous chapter, we will add MQTT connectivity and make use of third-party MQTT clients to receive and display the data coming from your prototype.

Chapter 9, *Presenting Your Own Prototype*, introduces technologies such as laser cutting and 3D printing to build great-looking and sturdy cases. It will give you an idea of how professional prototypes are made using **Printed Circuit Board** (**PCB**) services and software. While making a product out of your ideas probably seems far-fetched to you, I want you to explore different options and where continuing on this path might lead you.

To get the most out of this book

To get the most out of this book, you'll need to understand basic programming concepts such as variables, conditions, loops, and functions. You might have gained this knowledge by reading an Arduino book for complete beginners before that explained those concepts in detail. There are many books on getting started with Arduino for you to choose from.

If you run into any problems that you cannot solve yourself, feel free to ask for help on the book's repository on GitHub (https://github.com/PacktPublishing/Hands-on-IOT-with-MQTT) by opening an issue, and I will try to help you. To do so, you will need to create a GitHub account (if you do not already have one).

You can find all the required pieces of software, along with instructions on how to install them, in Chapter 4, *Setting up a Lab Environment*. If you skip the first hands-on project in Chapter 5, *Building Your Own Automatic Pet Food Dispenser*, starting directly with the second or third one (in Chapters 6, *Building a Smart E-Ink To-Do List*, and Chapter 7, *Building a Smart Productivity Cube, Part 1 and* Chapter 8, *Building a Smart Productivity Cube, Part 2*), please make sure to also follow the instructions on how to set up MQTT and the Wi-Fi library at the beginning of Chapter 5, *Building Your Own Automatic Pet Food Dispenser*.

You will get the most out of this book by reading Chapter 1, *The Internet of Things in a Nutshell*, to Chapter 5, *Building Your Own Automatic Pet Food Dispenser*, in a linear way. Chapter 6, *Building a Smart E-Ink To-Do List*, to Chapter 9, *Presenting Your Own Prototype*, can be read in any order you like.

Download the example code files

You can download the example code files for this book from your account at www.packt.com. If you purchased this book elsewhere, you can visit www.packt.com/support and register to have the files emailed directly to you.

You can download the code files by following these steps:

1. Log in or register at www.packt.com.
2. Select the **Support** tab.
3. Click on **Code Downloads**.
4. Enter the name of the book in the **Search** box and follow the onscreen instructions.

Once the file is downloaded, please make sure that you unzip or extract the folder using the latest version of:

- WinRAR/7-Zip for Windows
- Zipeg/iZip/UnRarX for macOS
- 7-Zip/PeaZip for Linux

The code bundle for the book is also hosted on GitHub at `https://github.com/PacktPublishing/Hands-On-Internet-of-Things-with-MQTT`. In case there is an update to the code, it will be updated on the existing GitHub repository. If you want to make sure to have the latest code you should download it from GitHub.

We also have other code bundles from our rich catalog of books and videos available at `https://github.com/PacktPublishing/`. Check them out!

Download the color images

We also provide a PDF file that has color images of the screenshots/diagrams used in this book. You can download it here: `https://static.packt-cdn.com/downloads/9781789341782_ColorImages.pdf`.

Code in Action

Visit the following link to check out videos of the code being run:
`http://bit.ly/2oSjufZ`

Conventions used

There are a number of text conventions used throughout this book.

`CodeInText`: Indicates code words in text, database table names, folder names, filenames, file extensions, pathnames, dummy URLs, user input, and Twitter handles. Here is an example: "So, often, it is better to rewrite `delay()` calls to use a custom timer instead of using the `millis()` function."

A block of code is set as follows:

```
if temperature < 5°C {
    send me a reminder to use beanie and gloves
}
```

When we wish to draw your attention to a particular part of a code block, the relevant lines or items are set in bold:

```
if (inputValue == 1) {
    lastTimeOpenend = millis();
    isOpen = true;
    myservo.write(90);
}
```

Any command-line input or output is written as follows:

```
mosquitto_pub
```

Bold: Indicates a new term, an important word, or words that you see onscreen. For example, words in menus or dialog boxes appear in the text like this. Here is an example: "Click on **Tools** | **Serial Monitor** to open the serial monitor."

Warnings or important notes appear like this.

Tips and tricks appear like this.

Get in touch

Feedback from our readers is always welcome.

General feedback: If you have questions about any aspect of this book, mention the book title in the subject of your message and email us at customercare@packtpub.com.

Errata: Although we have taken every care to ensure the accuracy of our content, mistakes do happen. If you have found a mistake in this book, we would be grateful if you would report this to us. Please visit www.packt.com/submit-errata, selecting your book, clicking on the Errata Submission Form link, and entering the details.

Piracy: If you come across any illegal copies of our works in any form on the internet, we would be grateful if you would provide us with the location address or website name. Please contact us at copyright@packt.com with a link to the material.

If you are interested in becoming an author: If there is a topic that you have expertise in and you are interested in either writing or contributing to a book, please visit authors.packtpub.com.

Reviews

Please leave a review. Once you have read and used this book, why not leave a review on the site that you purchased it from? Potential readers can then see and use your unbiased opinion to make purchase decisions, we at Packt can understand what you think about our products, and our authors can see your feedback on their book. Thank you!

For more information about Packt, please visit packt.com.

Section 1: An Introduction to IoT and MQTT

This section offers a quick introduction to what the **Internet of Things** (**IoT**) is, why readers should care about it, what MQTT is, and how MQTT can help readers to build their own prototypes.

The following chapters will be covered in this section:

- Chapter 1, *The Internet of Things in a Nutshell*
- Chapter 2, *Basic Architecture of an IoT Prototype*
- Chapter 3, *Getting Started with MQTT*

The Internet of Things in a Nutshell

1

Kids who grew up in the 1990s, like myself, will have noticed a lot of changes to our everyday lives over the last 30 years because of the internet and its omnipresence. Having a wireless connection wherever we go, checking emails and Facebook on our smartphones: all of this didn't exist in the 1990s. Smartphones? We had mobile phones, yes, but with a monochrome display and an antenna, and they didn't have an internet connection. The multimedia highlight was playing the game Snake, a simple 2D game with a game logic that can be programmed in a day. There was no mobile internet and no wireless LAN. When we were visiting another city and got lost, we had to go inside a store and ask for help—no Google Maps.

Nowadays, the internet is ubiquitous and people are dependent on it for their day-to-day life. Everyone has a phone in their pocket, which they can use to find any information they want in seconds. According to a study by *Business Insider*, on average, there will be four internet-connected devices per human by 2020.

Smart devices come in various forms and shapes and are causing disruptions in every industry, just like the internet did in the early 2000s. Companies such as Google and Samsung have been developing wearable devices, such as smartwatches and health trackers, automated robots for industrial applications, self-driving cars, smart buildings, and smart home devices, as well as early-warning systems for tsunamis or earthquakes.

In this chapter, we will explore some of the devices and technologies that are driving the internet-connected device revolution. We will have a closer look at the following topics:

- Exploring smart homes
- Exploring smart cars
- Exploring industry 4.0 / the Industrial Internet of Things
- Prototyping for the greater good
- What is a prototype?
- Voice control
- Why should you invest in the IoT?

Exploring smart homes

The most prominent category of **Internet of Things** (**IoT**) devices for end consumers is **smart home**—connected electronic devices that are used in your flat or house. They often replace traditional electronic devices by enhancing them with an internet connection and a way to control them digitally. In most cases, they offer a smartphone app or integration with a smart home app, such as Google Home, which makes it possible, for example, to turn the device on and off from your sofa, change its settings, and check its status.

According to a study by *Zion Market Research* (`https://www.zionmarketresearch.com/news/smart-home-market`), the global smart home market is about to reach around USD 53.45 billion by 2022, with big tech companies such as Samsung and Google being at the forefront. Let's look at some examples of these smart home devices:

- Smart fridges
- Smart door locks
- Smart thermostats
- Smart scales
- Smart lights
- Smart pet food dispensers

These are only a few of the smart home devices available today. The potential of such smart devices to influence our lives is huge.

Let's look at an example of a smart device—a smart fridge—and see what benefits it brings over a non-smart device.

The main problems people tend to have with a regular fridge are the following:

- Food goes to waste because you have no clear overview of expiration dates for each item.
- You run out of a certain food because you did not keep track of your supply.

Smart fridges keep track of your groceries and their expiration dates and warn you when you might run out of your favorite food. When opening the fridge, it could also present you with recommendations for what to cook with the available ingredients.

To minimize food waste, it might detect that you have, for example, tomatoes, curry sauce, and onions that will go bad soon. It might then automatically run an online search and present you with recipes that include these ingredients for you to prepare.

If it is connected to third-party shopping providers, it might also take care of ordering food automatically for you—for example, when your milk is running low, it might order a new bottle for you, which will then either be delivered to your doorstep or be ready for pickup at your local grocery store, together with other items that you are running low on.

While you cannot expect that level of comfort from a smart device just yet, more and more devices in our households will have similar features that make our lives easier.

How smart devices connect to the internet

In most cases, smart home devices need internet access to function correctly, which means that you must tell your device how it can connect to your local network via a network name and password.

When you first set up a new smart device in your household, the process often looks like this:

1. You connect the smart device to a power source.
2. The smart device goes into setup mode.
3. The smart device opens an ad hoc network (see `https://en.wikipedia.org/wiki/Wireless_ad_hoc_network`).
4. You connect to the network provided by the smart device via your smartphone.
5. On your smartphone, you visit a special website that is served by the smart device, and which is only accessible when you are connected to its network.
6. You specify the network username and password on the website.

7. The information is transmitted to the smart device.
8. The smart device will close the temporary ad hoc network and connect to your normal network via the username and password you provided.
9. The smart device can go online via your regular wireless network.

The very same technique can be used for creating IoT prototypes with microcontrollers, such as Arduino, Raspberry Pi, or Particle Photon, to create a convenient setup routine. We will not cover this technique in this book, but this is something to keep in mind if you're thinking about giving a prototype to friends and family or with batch production.

For example, if you were to bring your prototype to your friend's house, it would not be able to connect to the internet there, because it does not know the network name and password to connect with.

Without a setup routine like the one described previously, the source code of the device has to be modified each time the device is placed into a different environment to successfully connect to the other network.

Useful and unnecessary use cases

If you have ever lived in a single apartment without too much knowledge about how to separate your clothes when doing your laundry, you might have ended up with a bunch of pink shirts because a red sock found its way into your collection of white shirts that were to be washed. If you don't know what I am talking about, it is one of the mistakes a lot of single people make when using a washing machine. Washing a red sock with your white shirts is not a good idea, and might result in your shirts taking on the color of the sock, thereby becoming pink.

Other things can go wrong. You might do something wrong when setting the temperature, accidentally washing your favorite Norwegian Christmas sweater at 90 degrees, causing it to shrink. Now you might wear it while exposing your belly on cold summer days.

Chances are high that you don't want these things to happen.

Imagine that your shirts had a voice and that, when you asked them about their color and washing preferences, they answered, "*Hi, my color is light-blue and I enjoy being washed at 30 °C. I will get sick when you tumble dry me, so please don't!*". Now your washing machine could ask each and every textile lying in the washing machine drum about their washing preferences and set the program accordingly, or warn you that the red sock might better be off with the other red socks.

This can be made possible by using smart tags, such as **RFID** (short for **Radio-Frequency Identification**).

Many such use cases are about to bring value to our lives and make our lives easier. However, many companies forget about their product's original use case. One example is a smart light bulb that cannot be used as a light source when it is having a firmware update (`https://twitter.com/BalrogGameRoom/status/1036644958973960192`), which lasts for up to one hour. A firmware update probably does not need to be done very often, but if the light bulb is going through a firmware update in the evening, when it is dark, you will question their update policy and want to go back to using a normal light bulb.

I think updating the firmware on devices is important to keep the device secure and safe from hackers, but, in this case, the company producing these smart light bulbs could have spent more time developing a better update routine to improve the user experience.

Often, smart devices only work under perfect conditions, in this case, with updated firmware. Compared to non-smart devices (in this case, a normal light bulb), this is a huge step backward. A normal light bulb has one purpose: provide light when it has power. Either it works or it does not. Adding internet capabilities to a device might not always be the best decision, as it produces problems such as the one mentioned previously. The device should be created in a way that means that using its core feature is not disturbed if there's no internet connection or outdated firmware.

Many companies that develop smart devices overemphasize the smartness of their products and forget about their original use case, reducing the usability of the device, which was its main purpose. Not all things need to be smart, and if they are made smart, then their main feature should work even without internet access.

In another example, two security researchers discovered a vulnerability in a smart vacuum cleaner model—a robot equipped with a camera cleaning your flat autonomously. According to their study, it is not too hard to get access to the vacuum cleaner with admin privileges, allowing an attacker to misuse the vacuum cleaner as a 360º spying device. Definitely not what you signed up for when buying a smart device to save yourself some time. In this case, the manufacturer should have spent more time securing the device properly to make sure it is safe from hacker attacks.

There definitely is a need for autonomous vacuum-cleaning robots, smart light bulbs, and probably smart coffee machines, but when building new smart devices, you should always critically decide whether adding internet access to the device really adds something to its utility. If it does, then you should bear in mind the following questions when thinking about these aforementioned edge cases:

- What happens when the device is not able to reach the internet? Is it still usable?
- Can the device be updated? If so, when will it be updated? Does an update block its main functionality?
- Is the device secure enough? Or will it be easy for hackers to access the device and use it for their purposes?

If you want a few hours of entertainment (and education), you should check out the Twitter account `https://twitter.com/internetofshit`, which collects IoT fails and available smart devices of little use. It will not only bring a smile to your face but also make you realize that things went too far for some companies when dealing with smart home devices. The first question to ask yourself should always be: *Is the thing I am about to build useful?* Just making something smart certainly does not make it more useful, but instead might make it vulnerable to hacker attacks or completely useless when the internet connection does not work.

Exploring smart cars

Smart cars, another emerging field associated with IoT, is gaining momentum. Its progress is closely connected to the advancements made in machine learning in the last decade. If you have never heard about machine learning, you should put this book aside for a second and watch the TED talk *The Rise of Artificial Intelligence through Deep Learning* by Yoshua Bengio (`https://www.youtube.com/watch?v=uawLjkSI7Mo`). In a nutshell, machine learning makes it possible for computers to learn just like our brain does. It is another future technology that will be paired with IoT more and more to create smart, self-learning devices.

Machine learning is used in smart cars to develop many of its essential features:

- Detect the street, other cars, and people
- Understand signs and speed limits
- Identify dangerous situations and know how to solve them (for example, by applying the brakes)

The following screenshot shows a simplified view of object detection in a smart car:

A simplified view of object detection in a smart car (image based on photo by Josh Sorenson)
Source : (https://www.pexels.com/photo/architectural-design-asphalt-buildings-city-139303/)

If you compare the visual interpretation of a human's view of a street and the digital representation of the same theme, the two differ enormously—a computer seeing through a camera just sees raw data, the amount of red, green, and blue per pixel, and nothing more. Machine learning makes this data more usable by training the computer based on input footage—for example, by supplying a large number of images depicting street views.

After many learning iterations, the computer might be able to identify a street using fresh footage. Machine learning and IoT will be good friends in the future as internet-enabled microcontrollers become smaller and more powerful.

Currently, complex machine learning models require an expensive state-of-the-art computer, but there are already experiments using the Raspberry Pi, a tiny computer that runs Linux, for simple machine learning tasks. Google and NVIDIA introduced two new development boards (so-called edge devices) in 2019, which have a similar form factor to the Raspberry Pi and are intended for machine learning prototyping: Google Coral and NVIDIA Jetson TX2.

But so-called on-device training is not the only way hardware devices can use the power of machine learning. The most common way they use machine learning today is by sending the device's data to a cloud server where the heavy analysis is done. One example of this is Google Photos. It allows you to upload your photos, in most cases taken with a smartphone, to the Google servers. The servers will analyze each and every one of them using various machine learning models.

You can already use the results in two ways, as:

- The machine learning model detects all faces in a photo and groups them together into categories. This way you can easily find all photos that contain your face or any of your friends' faces.
- The machine learning model detects objects in photos. You can then, for example, filter all of your images that contain a red car.

Apart from machine learning, to understand their surroundings, smart cars can communicate with each other. Every now and then there are reports about a mass crash on the highway. Smart cars will be able to warn each other about dangerous situations: *"Attention cars behind me. There is an obstacle lying on the street. Better slow down!"*.

When it comes to situations like this, the amount of time it takes for another car to receive this information can make the difference between life and death—one or even multiple seconds response time is just not good enough here, the response time needs to be in the milliseconds. If the cars were using the internet for communication, it might take too long. When sending data to a satellite and spreading it from there to all of the nearby cars, there are too many things that can go wrong and prevent the warning from being delivered in time. Fortunately, there is a solution for this: using a technology called **Vehicle to Vehicle (V2V)**. With this technology, cars can talk directly to each other by opening a network themselves (like a router). Using this, they create a mesh of connected cars without needing internet access.

Exploring industry 4.0 / the Industrial Internet of Things

Industry 4.0 refers to the fourth industrial revolution. One of its driving technologies is IoT, connecting physical machines digitally with each other and the cloud.

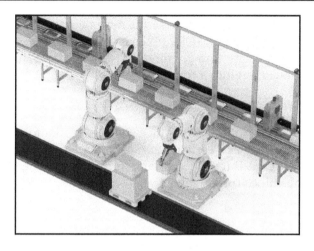

Industrial robots (source: Depositphotos)

According to *IoT Fundamentals: Networking Technologies, Protocols, and Use Cases for the Internet of Things*, by *David Hanes (et al.)*, (2017, Cisco Press), p 932-933, 1208-1209:

> *"There are estimates that there are 60 million machines in factories; the vast majority of them are more than 15 years old and 90% are not connected to the internet."*

In the airplane industry, preflight security checks involved a lot of manual work—each and every piece of essential equipment had to be checked off a list for the flight to be considered safe. On average, this took 6.5 hours per plane.

By integrating RFID tags into essential safety equipment, for example, the airplane industry made it possible for security staff to use an RFID scanner instead of a paper checklist to make sure that no important parts are missing.

And adding these chips was worth it. The 6.5 hours it took to manually check for the presence of each part could be reduced to 20 minutes this way.

There are many more industries less modern than the airplane industry that can profit from IoT as well.

By 2016, approximately 20 million smart meters had been installed globally, which can send their data automatically to the cloud. One of the areas where this is being used is power consumption in apartments.

Using smart meters and a web interface, tenants can check their monthly energy consumption. This way, it is easier to identify electronics or usage patterns that require a lot of energy, just by comparing the monthly costs.

Without smart meters, it is hard to tell whether the new electric grill you just bought is actually a power hog. You can imagine how much easier it is for the power companies as well: no need to send a technician from apartment to apartment, writing down numbers on paper that then have to be typed into a computer later.

Smart meters make everything much easier.

One of the areas where a lot of man-hours can be saved is semiautomatic maintenance of machines. Every machine part has a certain life expectancy: some fragile parts may last a few weeks, some several years. But sooner or later, physical parts need to be replaced. Most factories rely on their machines running in parallel—once one of them stops, the flow cannot go on.

Detecting machine parts that need to be replaced before they actually break can save the company a lot of analysis work and money. With self-monitoring machines, this is about to become more efficient and involve less human maintenance work. By equipping machines with various sensors to run self-tests and verify that all of their parts operate as planned, malfunctions or old parts can be identified early on by the machine. It can then call for a technician to replace part x or manually check part y. By pinpointing possible problems this way, machine downtime can be minimized.

The same self-analysis functions have found their way into the consumer market more and more as well. Commercial coffee machines have an internal counter that is incremented every time a coffee is made. After x coffees have been made, the machine might blink an LED to prompt you to run a manual maintenance program, for example, to get rid of unwanted deposits. While this isn't too smart, modern consumer 3D printers actually are. Being equipped with a multitude of sensors, they can detect malfunctions, identify broken parts, and fine-tune their own settings while printing.

Prototyping for the greater good

Building IoT devices can also contribute to the greater good in a non-commercial setting. In 2018, a non-governmental organization consisting of engineers and developers called **Rainforest Connection** (`https://rfcx.org/`) developed an IoT device to help to protect the Amazon rainforest from illegal deforestation. As it reduces the levels of greenhouse gases in the atmosphere, the rainforest plays an important role in our climate, and, according to Rainforest Connection's CEO Topher White, illegal deforestation accounts for nearly one-fifth of all the greenhouse gas emissions every year. Up to 90% of the deforestation of the Amazon rainforest is done illegally. Saving the rainforest could be the cheapest and fastest way to slow down climate change:

African jungle (source: Depositphotos)

In collaboration with the indigenous Tembé tribe, Rainforest Connection developed a device called **Guardian** to detect illegal deforestation and prevent it from happening. The devices are based on recycled smartphones, which have many of the ingredients of an IoT device on-board—a microprocessor, sensors, and a way to communicate via the cell phone's network over the internet.

The Guardians are hung inside the trees, their microphone transmitting the ambient sound via the cell phone network 24/7, forming a huge grid of microphones. In the cloud, where all of this data is assembled, a machine learning model based on Google's open source TensorFlow library comes into play. It was trained to detect the sound of chainsaws and trucks used in deforestation operations. Whenever one of the microphones detects a sound like this, disturbing the natural sound collage, the machine learning algorithms can identify it and the relevant information, such as GPS data, can be forwarded to the authorities.

In March 2018, the Planet Guardians program was launched by Rainforest Connection. Students from Los Angeles helped to build new Guardian devices to be added to the grid, and it is expected that these will help to protect 100,000 acres of rainforest throughout the year 2020 (you can go to `https://www.blog.google/technology/ai/fight-against-illegal-deforestation-tensorflow/` and `https://www.prnewswire.com/news-releases/rainforest-connection-introduces-one-of-the-largest-programs-ever-launched-by-students-to-protect-the-worlds-rainforests-300617270.html` for more information).

Similarly, devices equipped with GPS and a sensor are used to detect earthquakes—once an earthquake is detected by the sensors, people can be warned accordingly.

The same principle can be used for other areas as well—as an early warning system for tsunamis or avalanches, for example.

What is a prototype?

Before starting to work on your prototype, it is important to understand what a prototype is and what it is not. From a product perspective, various things need to be evaluated before a device can be produced in batches, starting with the functionality. Is the device useful? Does it serve a purpose? You also might want to find out how it feels. What material is it made of? Does it feel good when you hold it in your hands? How does it look? Do the buttons have a nice degree of resistance to pressure? Is it fitting in to its environment? Is it easy to use? Does it need a manual? Where can you get all the materials for batch-production?

Hardware prototyping (source: Envato Elements)

If your device is supposed to be outside, you also need to find out which materials can be used for it to survive the wind and rain. For all of these questions, you might want to produce one or many prototypes to get closer to a possible product, step by step.

A lot of the areas mentioned previously can be tackled individually—to find a nice form factor, you might design various models in a 3D application, print them out, and iterate upon them. The material used can also have a huge impact. By using online services such as Shapeways (`https://www.shapeways.com`), you can have a 3D printing service at hand that can print in various materials, such as plastic, aluminum, and gold. That last one might be a bit expensive, but at least you have the option. Combined with other rapid prototyping techniques, such as CNC milling or laser cutting, the possibilities are endless.

While building the prototype, it is essential to find out quickly whether your idea works, both from a usability perspective and from a technical perspective.

To evaluate the usability and usefulness of the device, you might not need to tinker at all. When developing software interfaces—for example, a design for an application—it is common to create so-called paper prototypes—a piece of paper for each state of the interface:

A paper prototype is used by designers to evaluate their idea as fast as possible (source: Depositphotos)

When evaluating whether the interface is working, a guinea pig needs to simulate using the interface, for example, by pointing on the piece of paper and saying *"I'm pressing the blue button"*.

The person testing the interface can then exchange the piece of paper for another one showing the resultant screen, as if a button was clicked on a working prototype on the computer, to show the new state. By doing this, many problems will be uncovered before a line of code is written. The same principle might be used for hardware prototyping. Let's say that you have a device with a button and an LED that can either turn red, green, or blue, depending on how often you press the button. If you actually draw a sketch of this prototype on paper, one page for each state (for example, one page where the LED is red, one where it is green, and one where it is blue) and simulate the usage, you might notice that, when pressing the button, you cannot see the LED anymore because your hand is covering the LED when pressing the button.

Hooray, you just discovered a usability problem that might have cost you hours of work if you had started soldering right away. Prototyping 3D objects with 2D paper is definitely not optimal, but it might save you some time anyway. The key is to evaluate multiple parts of your design fast to find out what works and what needs to be developed.

This is equally true from a technical perspective. Try breaking down your problem into multiple small problems to evaluate each one individually, especially when you don't feel comfortable with electronics. Let's say you want to build a device that reads a sensor value, and, when a button is pressed, tweet its value. You could build three mini sketches here to make sure each of the needed functionalities works individually:

- Sketch #1 might output a button press. With this sketch, you can verify that you connected the button in the right way and that its value is correct (pressed or not pressed).
- Sketch #2 might output the value of the sensor to the console. Interact with the sensor and find out whether the value is what you expected it to be. When using a distance sensor, it might actually only work in a very specific range and would not at all be usable for your idea. Hooray again, you saved yourself some time by failing fast, instead of spending hours on something that was always going to end up in the trash.
- Sketch #3 might just send a tweet.
- Sketch #4 might then combine all of the sketches to create the actual prototype you are after. When working on prototypes, I like to keep loose ends to a minimum. By working this way and validating different parts of your prototype early and in an isolated fashion, you will be more comfortable bringing it all together, and possible problems will be easier to pinpoint.

A sketch is a program that is uploaded to a microcontroller, such as an Arduino microcontroller. It is a widely used term and is used interchangeably to refer to a program or code snippet. It is a reference to sketches made by an artist and is used in Processing, the graphical development environment similar to Arduino (`https://www.processing.org`).

The beautiful thing about prototyping is that you don't need to be an expert in every area—you will find code snippets and diagrams for most of the common types of sensors and actuators around the internet, especially in the Arduino community. Combining example snippets and adding a little bit of logic to the code might do the job for a first version.

When prototyping for IoT, a main ingredient is connectivity. When starting new projects, making sure that the microcontroller can connect to the internet is one of the first things I do.

One of the areas that clearly goes beyond a prototype for IoT is security. Making sure your device is well protected against hacker attacks is another issue to be tackled once you actually finish your prototype. If you think about batch-producing a device, this is a topic that needs to be addressed and well researched so that your product isn't hacked and doesn't end up on the Twitter account `https://twitter.com/internetofshit`.

Voice control

With Apple's Siri, Microsoft's Cortana, Amazon's Alexa, and Google's Assistant, smart voice interfaces have found their way onto each of our smartphones, and they are about to be found in more and more gadgets in our homes:

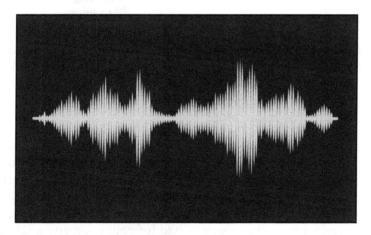

Waveform (source: Envato Elements)

The newest generation of smart speakers brings the vocal assistant as a key feature, not only allowing us to listen to our favorite tracks, streamed from Spotify via Bluetooth from our smartphones, but also acting as an interface to the web.

Using voice assistants to get the weather forecast for the day, find out what the capital of Finland is, or create a reminder to buy the milk makes these tasks a lot easier.

Natural voice interfaces open up a door to the connected world to control our homes and gain easy access to information. The technology is still young and full of flaws—many questions that you might ask the digital assistant are not understood yet, and result in an answer like "*Sorry, I could not understand you. Did you mean…?*". The complexity of conversations is also far off from resembling natural conversations between humans. Asking one of the voice assistants "*How is the weather tomorrow?*" is an easy task and the response will most likely be helpful to you and will let you know whether you should pack an umbrella or not. Asking a follow-up question such as "*And the day after?*" is a completely different problem and has not been supported until very recently.

Because of advancements in machine learning, these problems are about to be tackled, and the assistants get better each year, respond to more complex questions (and follow-up questions), and behave more and more like humans.

You might be wondering what all this has to do with IoT. When working on IoT projects, and especially wearables, you might have strict spacial restrictions. Let's say you want to create an LED necklace that can change color—being able to change the color would require physical buttons, which would take up extra space. It would also often feel hard to use without adding a display. This could already be too much to carry around. Using a voice assistant for this case would feel way more natural. The necklace could have one button to activate the assistant; once it is pressed, you could tell your necklace to change its color to blue, for example, using an external voice recognition service such as the Google Speech-to-Text API. From the API, you would then get the text *blue* back, which you could use in your Arduino code to actually switch the color.

Using external services and premade modules is one of the things I really want to push you toward. Creating a prototype is not about creating a consumer-grade product, ready to be produced in a batch of 50,000 in a factory in Shenzhen and hand-coded in tens of thousands of lines of code. Creating a prototype is about succeeding (or failing) fast in creating a functional prototype, either because you really want to build your own smart coffee maker or you want to find out whether that idea you had for your company might actually help with digitalization. Use whatever is available, hack it together. If one of the components is a proprietary voice recognition library, it is alright. If your prototype is doing well and you are thinking about bringing it to the next stage and actually producing it, you can still look for alternatives.

Why should you invest in IoT?

For one, working on hardware projects can be very fulfilling. In contrast to software, hardware prototypes are physical. They can be looked at from multiple viewpoints, touched, and taken with you. There is something magical about seeing your creation work on an actual device, and this is even more true with internet-enabled devices. Building a device that interacts with another device wirelessly is magical × 2.

Artificial intelligence (**AI**)—deep learning, image recognition, natural-language processing, and neural network-driven decision-making are advancing every year, and the possibilities of combining IoT with AI open up completely new possibilities (for more information, see `https://www.forbes.com/sites/bernardmarr/2018/01/04/the-internet-of-things-iot-will-be-massive-in-2018-here-are-the-4-predictions-from-ibm/#62a82b48edd3`).

IoT, with all of its devices, is connected closely to the concept of big data. Many companies want to analyze all of the sensor data stored in the cloud to draw conclusions from it that can then be used to maximize their profits. Here, machine learning comes into play to make use of the data and create rules to act upon.

Combined with modern frontend tooling using HTML, CSS, and JavaScript, it is possible to create, in just a couple of weeks, a working prototype of a sensing, internet-connected device in a nice case that analyzes its data using machine learning and presents its results and controls in a nice-looking web dashboard. Ten years ago, you would have needed a large budget and a lot of manpower for this task; now it can be done by only one creative technologist.

Summary

We now have a fair idea about the most important areas related to the Internet of Things.

One thing that all IoT devices have in common is their connectivity—equipped with an internet connection, they can send and receive information, sense their surroundings, collect data, and control physical actuators. To be available over the internet, these devices need to have unique identifiers, usually in the form of their IP addresses.

You learned that IoT is being used in smart homes in the form of connected devices (for example, smart fridges and smart heating systems). This area has the most possibilities for your own experiments using an Arduino, sensors, and actuators.

We also had a look at two other areas where IoT is being used that gain a lot of traction: smart cars and industrial IoT. Smart cars will take over our streets and (at least partly) replace regular cars. The industrial IoT will help to automate factories around the world, improve efficiency, and minimize the need for manual maintenance.

You learned how a prototype differs from a product and that building quick-and-dirty experiments is often the fastest way to validate your ideas.

You also got a glimpse of voice control and learned that technologies such as Siri and Cortana can be used in physical prototypes as well.

While many of the use cases of IoT are in the commercial space, there is also a lot of room to create prototypes for the greater good using IoT development boards as you learned in the section about the Guardian, a device to help in stopping illegal deforestation.

In `Chapter 2`, *Basic Architecture of an IoT Prototype*, you will learn more about the IoT ecosystem—microcontrollers, protocols, apps, and libraries.

Questions

1. Name three areas related to IoT.
2. Name one user interface trend for internet-connected devices.
3. Does it make sense to make every gadget smart?
4. What is a prototype?
5. Do you need to be an expert to create IoT prototypes?

Further reading

- **A study by Business Insider**: https://www.businessinsider.de/there-will-be-34-billion-iot-devices-installed-on-earth-by-2020-2016-5
- **A Zion Market Research study**: https://www.zionmarketresearch.com/news/smart-home-market
- **Forbes IoT predictions**: https://www.forbes.com/sites/bernardmarr/2018/01/04/the-internet-of-things-iot-will-be-massive-in-2018-here-are-the-4-predictions-from-ibm/#2541d333edd3

- *IoT Fundamentals: Networking Technologies, Protocols, and Use Cases for the Internet of Things.* David Hanes, Gonzalo Salgueiro, Patrick Grossetete, Rob Barton, and Jerome Henry, (2017, Cisco Press), p 932-933, 1208-1209

- **Light bulb firmware update gone wrong**: https://twitter.com/BalrogGameRoom/status/1036644958973960192

- **Coffee machine requesting email address**: https://twitter.com/paulgrimes_/status/1031328090859814912

- **Smart fridge**: *Aitken, R. et al. (2014) Device and technology implications of the Internet of Things (ARM), VLSI 2014, June 2014*

- **Vulnerability in vacuum cleaner**: https://techcrunch.com/2018/07/19/vacuum-vulnerability-hack-diqee-positive-technologies/

- **The (Smart) Home of Your Dreams**: https://medium.com/iotforall/the-smart-home-of-your-dreams-82837822f53e

- **The fight against illegal deforestation with TensorFlow**: https://www.blog.google/technology/ai/fight-against-illegal-deforestation-tensorflow/

- **Project Planet Guardians**: https://www.prnewswire.com/news-releases/rainforest-connection-introduces-one-of-the-largest-programs-ever-launched-by-students-to-protect-the-worlds-rainforests-300617270.html

- **Rainforest Connection**: https://rfcx.org/

- **Zismos—earthquake early warning system**: https://www.zizmos.com

- http://avatech.com/

- **Examples of IoT gone wrong**: https://twitter.com/internetofshit

- **The Rise of Artificial Intelligence through Deep Learning by Yoshua Bengio**: https://www.youtube.com/watch?v=uawLjkSI7Mo

- **Shapeways—3D printing on demand**: https://www.shapeways.com

Basic Architecture of an IoT
Prototype
2

In this chapter, you will be introduced to the basic building blocks of an IoT project. You will briefly get to know various protocols, architectures, and development boards for IoT prototyping. The main purpose of this chapter is to show you what's out there, so you can decide which technologies might be of interest for your next project and give you a head start on further exploration. If you feel overwhelmed by the number of new technologies and options available, don't be. You don't need to understand everything here, but hopefully you will come back later to look up one or more topics if you stumble upon them elsewhere. This chapter neither tries to be a complete reference, nor can it be. It would go completely beyond the scope of this book—which is to teach you how to build IoT prototypes with Arduino and MQTT. After reading this chapter, you will have a better idea of technologies relevant to IoT prototyping. For many of the topics mentioned here, you will also find (at the end of this chapter) additional articles or videos that go into more detail.

The following topics will be covered in this chapter:

- Building blocks of IoT connectivity
- Understanding protocols and communication
- Exploring microcontrollers for IoT

Building blocks of IoT connectivity

One thing all IoT devices have in common is connectivity. In most cases, this means that the device can connect to the internet; in some cases, it might communicate with a local smart hub via another technology such as Bluetooth, which is connected to the internet itself, and sometimes devices from a local network, for example, via radio (see the *ZigBee* and *Thread* sections). In each case, there is a communication channel, so devices can either send out data, receive commands, or both.

Having a device that is connected to the internet or another device is much use. It needs to be connected to input and output components. With sensors as input components alone, there are a gazillion possibilities. If you haven't done so already, you should have a look at the sensor category for Sparkfun, one of the best-known shops (`https://www.sparkfun.com/categories/23`). Most sensors to be found there are easy to integrate and come with a hookup guide and an example Arduino sketch to get you started quickly. Reading the user reviews there will also give you a good idea as to whether a certain sensor performs well. If you find yourself looking for new ideas about what to build next, browsing the available sensors might give you some inspiration.

The same is true for output modules—physical components such as motors, screens, speakers, or lasers. You could, for example, connect a motor and make it spin, display information via LEDs or a screen, use solenoids—a special form of electromagnet—to create an electronic door lock, or create a smart watering system using a water pump. For your device to interact with other devices (for example, a web server) on the internet, your device needs to be identifiable. Humans have names. If I am in a room with five people, each of us with a different name, Lisa will know that I am talking to her when calling her by name. Let's say I could scream very loudly and everybody in my city would hear me—screaming "*Lisaaaaaa!*" would be problematic, because there are 16,923 Lisas living in my city. To address her in a bigger network, I need a unique identifier, for example, an email address.

For smart devices, the unique identifier is often its IP address, a number such as `123.45.123.45`, which is unique across the globe and is automatically assigned by your internet provider. If your smart device interacts with a web server, such as "*Hey weather.com, smart mirror here, please tell me how hot it will be today in Berlin, so I can display it*", the server needs the IP address of your device to send back the information your device is requesting: "*Sure buddy, it will be 23° today, but pack the umbrella!*".

In addition to the IP address, which might be different each time your device goes online, you probably want to interact with another server. In our case, this will mostly be a third-party server used to store and exchange data, something we gladly do not have to code ourselves when prototyping our smart device. There are many ways to store information for free on the internet; for example, we could create a spreadsheet document in Google Drive and write down our sensor data there. But there are better options, specifically designed for machines communicating with other machines, on your custom-made smart devices. These services offer an **Application Programming Interface (API)** for your device to interact with. Often, these APIs use the **CRUD** pattern, which stands for **Create**, **Read**, **Update**, and **Delete**. When working on Arduino projects, two kinds of third-party web services are particularly relevant to us: data storage solutions and webservices to visualize your data. When working on hardware projects that collect sensor data, you need a place to store it. In many cases you want to store this information using a third party webservice that puts your data in a database and offers a way for you to access it, often in a tabular view, where you can inspect the raw data or download it. Here, visualization webservices come into play—they allow you to import your data and visualize it using graphs. This allows you to make sense of your data and visually inspect it, which often reveals hidden information that would remain uncovered when just looking at the raw data.

Some webservices combine another two features: they offer a way for you to store your sensor data and, at the same time, visualize it.

Programming web servers is not easy, so we can be happy that this part is taken care of for our prototypes and we don't have to think too much about it: we just have to choose one to use. There are many options available today for cloud services targeted at makers and hardware start-ups. Within a day, you can try them out and decide for yourself which one you like the most. In general, I would stick with those targeted at makers and try to avoid the big players such as AWS and Google Cloud, because they are targeted at big companies and add a lot of complexity to your projects.

More and more microcontroller manufactures have their own cloud service for you to log sensor data. For example, Particle, a leading manufacturer of microcontrollers for IoT, has its own cloud service called Device Cloud (`https://www.particle.io/device-cloud/`). Adafruit, another manufacturer of microcontrollers, as well as a web shop for makers, offers Adafruit IO (`https://io.adafruit.com`). Similar to the Particle cloud, it can be used to log data and visualize it in a web-based dashboard. Adafruit IO is compatible with MQTT, so you could use it as a place to store your MQTT-powered projects' data. Another handy feature of Adafruit IO is the possibility to set up notifications via triggers. Most of these services are logic-less though, meaning that you can store data, for example, temperature: 5°C, but you cannot create rules as follows:

```
if temperature < 5°C {
    send me a reminder to use beanie and gloves
}
```

For logic to act upon your data, you either need to create rules on your device itself or integrate a third-party service such as **IFTTT (If This Then That**, `https://ifttt.com`). Here, it is really easy to create rules without writing a line of code. The rules you create can utilize a multitude of other services such as Facebook or Twitter. Creating a rule to send out a tweet such as *"Pack your sunglasses"* when the temperature sensed on your device is over 30 °C is done in the blink of an eye. IFTTT offers various channels; one of them is named `Webhooks`, which can be used to communicate with a microcontroller (see `https://ifttt.com/maker_webhooks`).

Mostly, these cloud services come with a generous free tier. As a hobbyist building a prototype, the free-plan limits are more than enough. If you decide to go into production and stick to the same web service, you will have to pay at some point.

Client-server architecture

To see the possibilities of IoT prototyping, it is important to understand the client-server architecture. When you visit a website in your browser, let's say, `https://www.arduino.cc`, your browser makes a request to the web server, which can be found via the domain, `arduino.cc`. The server will then send an HTML file back to the browser. Once the HTML file reaches your browser, it will start interpreting the file and look for other files that are needed to display the page properly. This might be images, videos, style sheets, web fonts, or additional scripts. Your browser will then display the website. This is called the frontend. The frontend is the part of a website that runs in your browser, after the files have been requested by the backend. The backend can not only send files back, it might also have access to a database that can be used to store sensor data:

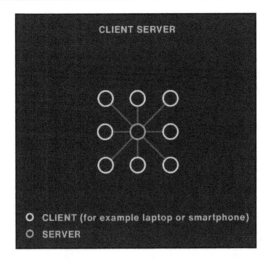

Client server architecture: Multiple clients send and receive information to one server

A client can be basically any device that is connected to the internet—smartphone, tablet, notebook, custom Arduino device, or another smart device.

Web interface

To control your IoT device, you will need some sort of interface that can either be a website, smartphone app, or desktop app. Many smart home devices are shipped with their own app that allows you to control the device remotely, change its settings, or check its sensor values even when you are on the road. With recent developments in frontend tooling, it is now easier than ever to create web applications, desktop apps, or mobile apps using web technologies. If you are comfortable using HTML, CSS, and JavaScript, you can use one of the available frameworks to bundle a normal website as a smartphone app, a desktop app, or both. Using a third-party service as a data store, such as those we talked about before, this can be done in a couple of days.

The most common framework to build cross-platform desktop applications using web technologies is Electron (`https://electronjs.org/`). For mobile app development using web technologies, there are even more good options. Cordova (`https://cordova.apache.org/`) is one of the most famous. Similar to Electron, it allows you to create an app, using web technologies, that runs on Android and iOS and which can communicate with your smart device, by using just HTML, CSS, and JavaScript. Some other frameworks worth mentioning here are PhoneGap (`https://phonegap.com/`), Google Flutter (`https://flutter.io/`), Ionic (`https://ionicframework.com/`), and React Native (`https://facebook.github.io/react-native/`).

If you are not planning to build an app as a companion for your IoT device, you don't have to remember any of these technologies mentioned previously. Using these frameworks and tools is not necessary to follow the practical part of this book, and they are just named here to give you a head start if you decide to dig deeper. But before working with any of these frameworks, I would recommend spending some time learning JavaScript properly. It will make your life a lot easier later on.

Application Programming Interface

APIs glue the web together. They are a way for computers to communicate with one another. By browsing the web, you hardly notice the existence of APIs, but they are everywhere. If you have ever searched for a flight online, you probably stumbled upon one of these websites claiming to offer the cheapest flights around. You enter where you want to go and on what day and, in the background, it will run a lot of API requests to the servers of different airline companies, compare results, and then present you with the cheapest options it could find.

Many websites offer two ways to access their information. The first one is for humans, like you and me: it displays the information in a web interface with a **graphical user interface (GUI)**. It has buttons, often a search bar, big headings, and hero images to guide your eyes. If a machine needs to access this information—in this case, the weather—this is not the optimal way to go. For a machine, it is hard to figure out where a search bar is, which button to click after the search query has been entered, and where the desired information is placed on the result page once the search has completed. Finding out what the temperature in Berlin will be the next day this way is a hard problem for a machine.

Imagine you are asking for directions in a foreign country and you just cannot understand their language. Web frontends are not meant to be interacted with by machines. For this, there are APIs, a way for machines to talk to one another. Each API has its own definition, which the developers of the API decided upon and which developers of other software, who want to interact with this API, can read up on in the API documentation.

Every time we humans interact with each other in a formal manner, we use the same principle. Let's say we go to a restaurant. After entering, the waiter asks us how many seats we need and waits for us to respond with a number: 3. He will then guide us to a free table with 3+ seats, give us the menu, and later on, ask us what we would like to eat and drink. The waiter will pass on our order, let the cook know that we, for example, don't want any peas in our dish, and after some processing (cooking) time, he will then bring us what we ordered. After we are done eating, we have to call him again, ask for the bill, and pay. APIs work in a similar way. There are certain rules that we have to apply and certain terms we might use to get what we want (food) or to pass on something ourselves (pay the bill). The waiter here can be seen as the API that allows us to get food from the kitchen, acting as a middleman between us and the chef.

While building physical IoT prototypes, you will deal a lot with third-party APIs. You might interact with a logging service such as ThingSpeak (`https://thingspeak.com/`) to store your sensor data in the cloud; log special events, for example, when your device encountered a sensor peak, in Google Calendar; or receive the latest stock prices to be displayed in a smart mirror—all via APIs. On the other hand, you might want to create an API for your own smart device, so other machines are able to interact with it.

In a former project, I created a custom lamp using an LED strip that had various modes—normal lamp mode, blink, lava lamp, and disco. Each of these modes also had different options, for example, light color or animation speed. Using a custom MQTT API, the lamp was integrated into the server of the company's main product and hung in the office of their headquarters. The server constantly monitored user activity on the web platform and passed it on to the lamp to slightly change its color and serve as an ambient light. Via the lamp, the color of the room always reflected the activity on the server, so the employees always had a feeling for how many visitors were using the web platform just by interpreting the color of the lamp.

White represented normal activity, greenish-white more traffic, and reddish-light less traffic. To spot network intruders (for example, via port scanning, a technique often used by hackers), the server told the lamp to blink red whenever a possible intruder was detected.

In many cases, you want to store or receive information privately—only you should have access to your sensor data. For this, web services where you have a private user account offer a way to authenticate yourself, so the system knows it is really you who wants to access your data. Depending on the kind of API, authentication can be implemented in various ways. When communicating via MQTT, your device needs to authenticate once at the beginning and can then send and receive data without the need to authenticate again. When interacting with a RESTful API, which I will explain in the following sections, you mostly have to send a security token with every request.

Representational State Transfer Application Programming Interface

Now that you know what an API is, let's talk about a specific kind of API, the **Representational State Transfer** (**REST**) API. Many current web services use this architectural style for their APIs. While which underlying protocol needs to be used is not defined, in most cases, it will be HTTP—the same protocol you use when you surf the web with a web browser.

Using REST, resources on the web are modeled in an object-oriented way—each resource has a unique URL, which can be used together with one of the underlying HTTP methods (often called verbs) to perform stateless operations on resources. Stateless means that the server will not remember any of your prior requests—each request stands on its own. When surfing the web, you typically use two HTTP methods already on a daily basis, GET and POST. Every time you enter a website into your browser's navigation bar and hit the *Enter* key, you run a GET request with the specific URL: GET website at the URL: `https://google.com`.

A POST request is performed every time you send a form, for example, when signing up for a newsletter on your favorite website. When you press the **Submit** button, your email address is sent to the server as a POST request. POST is used to store new data. PUT and PATCH are used to modify data, for example, when you change your username on a social media platform. To tell a RESTful server to delete content, the DELETE method is used. To exchange information, **JavaScript Object Notation** (**JSON**) is commonly used, which is machine-readable and well supported in every major programming language.

It doesn't matter if this is a bit blurry to you right now. Many of the libraries you are about to use abstract this away for you. But you will probably stumble upon GET and POST in future projects. Just remember that GET is used to receive information, while POST is used to store data.

To try out an API, the fastest way is mostly to use your web browser to receive information. Remember: every time you visit a website in your browser, it performs a GET request. But let's actually look at an API endpoint and inspect the result. To get weather information for Berlin in a machine-readable form, we can use the OpenWeatherMap API.

Please note: the following API call is an example from the openweathermap website and does not return real data (you might spot London in the response data instead of Berlin). In order to get live data, you would need to register your own (free) app ID: `https://samples.openweathermap.org/data/2.5/weather?q=Berlinappid=b6907d289e10d714a6e88b30761fae22`.

When requesting the preceding URL in a web browser (and therefore running a GET request), we should get a response in JSON format like this:

```
{
  "coord": {
    "lon": -0.13,
    "lat": 51.51
  },
  "main": {
    "temp": 280.32,
    "pressure": 1012,
    "humidity": 81,
    "temp_min": 279.15,
    "temp_max": 281.15
  },
  ...
}
```

In this case, what we are probably interested in is the `temp` property in `main`, the temperature in Kelvin, which we may then convert into degrees Celsius, if this is your local temperature unit. But let's have another look at the URL we requested to get this response: `https://samples.openweathermap.org/data/2.5/weather?q=Berlinappid=b6907d289e10d714a6e88b30761fae22`.

There are a few things here that you should be aware of. `2.5` here is the API version. When developers make breaking changes to their API, for example, they correct a spelling error in the URL, they cannot just update their server with the new version. Other devices, which rely on the API, would stop working. Therefore, APIs often use a version number, in this case, `2.5`. When you integrate this API in your project, you can be sure that the API will not suddenly change. There might be a version 3 coming out at some point, but version `2.5` will continue to work, and so will your project.

After the version number in the URL, you see the word `weather`, describing the resource you requested, followed by the `?` character, which introduces multiple key-value pairs. Key-value pairs always come in the same format:

```
key1=value1&key2=value2&...
```

Here, two key-value pairs are used to pass additional information to the server—q and appid. q (for query) contain the search term/city we are looking for, and appid is an identifier for your account, which you have to sign up for on the OpenWeatherMap page. In this case, it is an example key from the OpenWeatherMap API documentation page for you to try out. This kind of authentication is needed for most RESTful APIs. Imagine a user accidentally making a never-ending loop, in which they request data from the server—using an API key (sometimes called App ID, like here). The API provider can detect that the same client runs too many requests, block this particular key, and send a warning response back, which might look like this:

```
{
    "error": "Too many requests! Only one request each second allowed."
}
```

Now that you have an idea how the frontend and backend play together and what an API is, let's have a look at some of the most important protocols for IoT communication.

Understanding protocols and communication

In this section, we will have a look at four common protocols and technologies used for communication between IoT devices. **ZigBee** and **Thread** are both technologies that use radio to transmit information. ZigBee has long been the preferred way to transmit information between devices locally (without using the internet, but using radio signals), but is about to be replaced by Thread.

CoAP and **MQTT**, on the other hand, communicate over the internet. They are both intended to be used with constrained devices and have different areas where each protocol shines. We won't go into too much detail about CoAP and will concentrate on MQTT, starting with Chapter 3, *Getting Started with MQTT*, instead.

ZigBee

ZigBee is a specification for local networks using low-power radio signals. It is used in home automation and sensor networks when multiple devices need to talk to one another in relatively close proximity (10-100 m line of sight) without an internet connection. Using ZigBee, mesh networks can be formed, bridging longer distances to transmit information via intermediate nodes. Each device in a ZigBee network needs to be set up as a coordinator, router, or end device. Each ZigBee network needs to have exactly one coordinator, which forms the root of the network. Router nodes are able to forward data between nodes, as well as consume and send data, while end notes consume or send data.

While there are various advantages, such as low power consumption, there are also trade-offs—one of them is that transmission speed is very low compared to Wi-Fi.

In the following screenshot, you can see various networks formed using ZigBee end devices, coordinators, and routers:

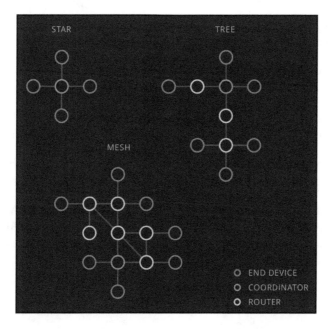

Devices in a ZigBee network can either be an end device, a coordinator or a router

You typically include ZigBee in your project by attaching a ZigBee module to your microcontroller. The most prominent ZigBee modules are called XBee and come in various forms and shapes. Once a ZigBee module is connected and (via third-party software) set up as one of the three node types (coordinator, router, or end node), it is easy to receive and transmit information.

CoAP

Constrained Application Protocol (**CoAP**) is an IoT protocol for machine-to-machine communication. It has its roots in the HTTP protocol, which we talked about before—HTTP is the queen of internet protocols and is used to access content via web browsers.

CoAP follows the RESTful architectural model we briefly talked about before as well—your device can have different resources, let's say, a temperature sensor and a humidity sensor. It may then be accessible via its URL:

```
coap://host:port/api/v1/temperature
coap://host:port/api/v1/humidity
```

Similarly to a RESTful API, you could receive its value by running a GET request to one of the preceding URLs. To store data, you would use POST.

CoAP and MQTT play in a similar league—they are both widely used, lightweight machine-to-machine protocols for the IoT, and both options have prominent supporters.

In this book, we won't use CoAP at all and will stick to MQTT instead, because it is currently better supported (more software libraries and tutorials) and is easier to use.

If you want to learn more about the pros and cons of MQTT versus CoAP, you should watch the *Comparison #144 Internet Protocols: CoAP vs MQTT* video, by *Andreas Spiess*, on YouTube (https://www.youtube.com/watch?v=pfG8uEDZj5g). His YouTube channel in general is highly recommended.

MQTT

MQ Telemetry Transport (**MQTT**), the protocol we will be using during this entire book, is a lightweight machine-to-machine protocol that makes it very easy to exchange messages between devices. The protocol uses a publish-subscribe pattern, which means that messages published by one device can be consumed by many devices. Each client can both send and receive messages.

There is another related protocol called MQTT-SN (for sensor networks). Unlike MQTT, it is used on non-TCP/IP networks (for example, ZigBee). We will not be using MQTT-SN in this book. If you are interested in this special version of MQTT, you can read more about it in its specification (`http://www.mqtt.org/new/wp-content/uploads/2009/06/MQTT-SN_spec_v1.2.pdf`).

We will discuss MQTT and all of its features in `Chapter 3`, *Getting Started with MQTT*.

Thread

Thread is a mesh networking specification for IoT devices, with impressive company backing. The Thread Group, which works on the definition of the protocol, consists of companies such as NEST (Google/Alphabet), Apple, and Samsung.

Similar to ZigBee, Thread is a wireless radio protocol that can be used to form mesh networks. Some key features of the Thread protocol are energy efficiency and security and the fact that it is based on IPv6, which makes it easier to address specific nodes, independent of the technology used to reach them (Wi-Fi/radio), both inside and outside the network. In comparison to ZigBee, creating node meshes is also easier with Thread, because you don't have to care about which node is a router and which is an end device. The Thread protocol will configure itself and select which nodes perform the routing. It also makes the network more resilient—whenever one of the nodes fails, it will self-heal and find another route for the traffic:

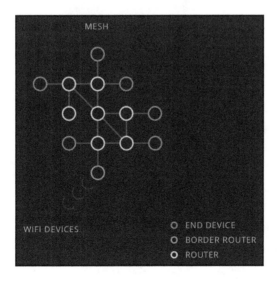

Mesh networks take care of the routing and device roles automatically

Think of Google Maps here—if you want to go from A to B, there are various routes to choose. If the fastest route is blocked because of construction work, it will present you with the second-fastest route. The same happens with Thread—router nodes are aware of the other nodes in the network and find the fastest way to route traffic, even if the nodes in between are malfunctioning.

While Thread is relatively new, microcontroller companies have started to adopt this protocol, the first one being Particle, which released three new development boards with OpenThread (`https://openthread.io/`)—the open source implementation of Thread—support in October 2018 (`https://www.particle.io/mesh/`).

Exploring microcontrollers for IoT

There are a multitude of microcontrollers available on the market today that are well suited for IoT prototyping. For simple projects where you just want to read a sensor value and log its value in the cloud, most microcontrollers with internet access will do a great job. In more complex projects, where you need a filesystem and more features than you would typically find on a regular computer, a microcontroller might not be enough, and you may need a development board running a full-grown operating system such as the Raspberry Pi (`https://www.raspberrypi.org`) or Asus Tinker (`https://www.asus.com/de/Single-Board-Computer/Tinker-Board/`):

Various development boards

These development boards come with display connectors and USB ports to connect hard drives, webcams, a keyboard, or a mouse to, and are able to run programs written in various programming languages. Besides giving you many options, this also adds a lot of overhead and makes development harder. You have to take care of certain things yourself. For example, when you write a program for the Raspberry Pi, you have to manually put it in the autostart for it to run when the device is rebooted. Arduino and co. automatically run the program, typically written in C++, on reboot. There are many other cases that will make your life a little bit harder when working with a full-grown development board using Linux or Windows 10 IoT Core (a special version of Microsoft Windows 10 for development boards such as the Raspberry Pi). On various occasions in my experience, GPIO ports were blocked by the operating system and could only be unblocked by a restart, something that never happened to me with a development board such as the Arduino.

If you don't require a screen, a mouse, a keyboard, a webcam, or speakers, you probably don't need a full-grown board running Linux for your next project and are better off with a development board programmed with the Arduino IDE, such as the one we will be using during this book—the Arduino MKR WiFi 1010 (`https://store.arduino.cc/usa/mkr-wifi-1010`).

The most important thing to look out for in a development board for IoT projects is connectivity. In most cases, this means internet access via a wireless network. Another approach would be a connection via Bluetooth to some kind of hub (for example, your computer), which is then connected to the internet and can therefore act as a bridge between the development board and the internet.

There might be cases where Wi-Fi or Bluetooth connectivity is not enough. Let's say you want to send sensor data from your development board to the cloud, but don't have a Wi-Fi network to use. This might be the case when your device runs on battery and cannot rely on the infrastructure supplying a network around it, for example, a prototype installed somewhere in the forest whose function is to sense air quality. Because these devices are not connected to a wireless Wi-Fi network (in this case, because there is none in the forest), other ways must be used to connect them to the outside world.

Cellular networks might be the answer here: the same kind of connection your smartphone uses to go online while you are on the road. Prominent boards for this use case are made by Particle (`https://www.particle.io`). The Particle Boron board comes with a cellular subscription, where you can use up to 3 MB of data each month (the amount of included traffic may change).

Another trend in IoT hardware is mesh networking. So, instead of only one connection to the cloud, hardware devices are connected to each other and the cloud. This has various advantages. If one of the devices loses its internet connection for some reason, it will still be reachable due to the mesh connectivity when within the range of another mesh device. Communication between mesh nodes is also faster, because the signal does not have to be sent to a satellite and back, but can be sent directly and therefore has to travel shorter distances.

3.3V versus 5V

There are currently two common voltages on which microcontrollers operate—3.3V and 5V. While, in the early days of the Arduino, 5V was the absolute standard, things are moving toward 3.3V and probably even lower voltages in the coming years.

Internally, a development board might work with different voltages. Many microcontrollers, for example, the Arduino MKR WiFi 1010, are powered with 5V, but all input and output ports can only tolerate up to 3.3V. This is something you have to look out for to avoid damaging the ports. Having a microcontroller that runs on 5V as the input voltage is handy because you might already have various USB power supplies or a battery pack to power smartphones or other USB devices at hand.

The important voltage here is not the input voltage, but what voltage the input and output ports accept. There are many sensors to be found that run only on 5V or 3.3V and not both. If the project you plan to work on uses a very specific sensor, its voltage requirements may help you to pick a development board. If the sensor is made to be run on 3.3V, choosing a development board that uses 3.3V as well will save you some time.

But what if you want to use a 3.3V development board and a 5V sensor together? For this, there are logic-level converters (also called level shifters), small components that are easy to add to your project and convert various 5V signals into 3.3V and vice versa. Sometimes, you may not need them though. If you're running a 3.3V microcontroller and using a sensor that requires 5V, it might run on 3.3V anyway (probably less reliably).

Over-the-air updates

Sometimes, it may happen that the code you've uploaded to a board needs to be updated, either because your initial code had some flaws and caused an error, or because you want to improve the code and add functionality.

Some development boards offer **over-the-air** (**OTA**) updates, a feature that comes in handy once your prototype is wrapped up, maybe boxed in a case, and where it is hard to physically connect to your computer via USB cable again in order to update the firmware.

Devices offering OTA updates do not need to be connected to your computer via cable; they just need to be connected to the internet and can then be updated from anywhere in the World. Currently, this is only supported out of the box by Particle development boards.

Boards using the ESP8266 chip (for example, NodeMCU) can be manually set up to be updated over the air as well, but it is not as easy and brings some security risks with it. If OTA updates are important to you and you want to spend the least amount of time setting it up, you probably want to use one of the available Particle development boards.

Open source hardware and clones

The embedded electronics market would not be the same today without companies such as Arduino or the Raspberry Pi Foundation, which release all or most of their hardware designs under an open license and push innovation forward. Boards licensed under an open hardware license can be studied, modified, and redistributed by anyone. Instead of treating a development board as an integral unit of your project and combining it with other modules, you can create a variation of the board, add or remove some components, and adapt it to your needs. Your variation of the board can even be sold without license fees. Arduino, for example, is licensed under a Creative Commons Attribution Share-Alike license. If you create your own version of one of their boards, the only requirement is that you have to give credit to the original creators and release your design under the same license as well.

 If you intend to do so, please study the specific board license in detail as there may be specific clauses.

Because of the open approach, there are many Arduino clones to be found, for example, on AliExpress, one of the biggest Chinese shopping portals. The names of the clones to be found there include, for example, the Nano 3.0 controller, compatible with Arduino, or the Nano 3.0 controller for Arduino, while the official one that can be found in the official Arduino store is named Arduino Nano (the version number, 3.0, is only mentioned in the product description). These clones can be purchased much more cheaply than their originals and mostly have a similar quality. There are differences though: some of the Arduino clones that you can find on the internet use an alternative Serial to USB chip (CH340), which might not work out of the box and may require extra drivers to be set up. Also, there may be slight variations in quality, because of variations in the hardware design, often to save costs in production. Overall, you will have fewer problems buying the originals.

The main reason to support the original developers such as Arduino or the Raspberry Pi Foundation is not better quality alone, but to support innovation. Somebody has to design the next generation of boards and push tools such as the Arduino IDE (editor) forward. If everybody were to buy cheap clones, newer models could no longer be developed.

Microcontroller board recommendations

To make your life a little bit easier while choosing a microcontroller board for your next project, I would like to introduce you to a few well-known ones, which are widely used and have different areas in which they shine. While we will use the Arduino MKR WiFi 1010 exclusively in the hands-on part of this book, it is good if you know what's out there and how the popular IoT boards compare to one another.

Particle Argon/Boron/Xenon

Particle is one of the major players in the IoT development space. Their first development board, called **Core**, offered an affordable way to prototype connected devices without needing to buy a Wi-Fi shield. Their second generation board, the Photon, brought prices down even more. Since November, 2018, their third generation of IoT development boards have been available and makes the Thread protocol available in a development board for the first time. Thread, which we discussed in an earlier section, is a protocol for mesh networking. So, having multiple devices that need to communicate with one another and rely solely on the availability of Wi-Fi is not an option; adding mesh to the mix might be the solution. It is also a great option if you have various devices that need to communicate with one another entirely without an internet connection. The following image shows a Particle Photon—the predecessor of the current generation of particle boards:

Particle Photon development board, predecessor to the Argon, Boron and Xenon

Particle boards can be programmed in two ways—either via the cloud editor on the company's website or via Particle Dev (`https://www.particle.io/developer-tools/`), a desktop IDE. It is not as convenient as the Arduino IDE, because you might run into little issues here and there, but overall it offers a good user experience. The library support for external modules, such as sensors and actuators, is not as good as with Arduino, but most are supported.

A great feature baked into all Particle boards is OTA updates. You don't need to connect the Particle development board to your computer if you want to flash new firmware; it just needs to be online somewhere. This means that the device could be in a different room, city, or country. As long as it is connected to the internet, you can upload new firmware. It's pretty handy! But it also has a (little) downside—when flashing new code, it takes longer than if you just flash your Arduino via a USB serial connection. It's not crazy long, but longer.

The third generation of Particle boards all offer support for the Thread protocol that allows it to form mesh networks. The board comes in three different versions, namely, Argon, Boron, and Xenon. The Particle Argon board comes with Wi-Fi support; the Particle Boron offers LTE, so it can interact with the internet via a cellular network; and the Particle Xenon can be used as either a node or repeater in a mesh network without Wi-Fi or LTE support at a smaller cost.

Via the Particle Device Cloud (`https://www.particle.io/device-cloud/`), you can manage all of your Particle development boards. One feature this platform offers that I find especially handy is checking the network status of each of your Particle devices. You can, for example, see which of your devices are currently online or offline, which often helps with remote debugging.

Another great feature of the Particle Device Cloud is the event dashboard—it makes it incredibly easy to log sensor data to. For example, if you have a temperature sensor attached to your board, you could upload its value every minute and then see the logged entries in the dashboard.

Particle also offers a smartphone app that you can (and need to) use before your device can connect to your local network via the Wi-Fi credentials. Via the app, you can also monitor the pins of each connected device and change their values, for example, to switch on and off an LED on the board.

Particle makes it extremely convenient to work on IoT projects, offering a complete suite of solutions, from prototyping to production. For hobbyists, using their cloud infrastructure for logging purposes is free of charge. Once you use it with more than 100 devices, you have to pay for it (these limits may be different by the time you read this).

I would definitely recommend that you get one of the Particle devices and use it for a project. While they offer a complete suite of solutions, you become dependent on their infrastructure though. Using a Particle device, you cannot easily move away from the Particle cloud. If you use their servers for logging and messaging and to flash the firmware, the device would be pretty much useless if their servers go down or they decide to charge for their services. I don't think this is likely to happen, but for this reason, in this book, we use more open approaches by using MQTT and Arduino. The Arduino can be flashed without the need for an external server, and MQTT is an open protocol with many public servers around. If the provider of the MQTT server you are using decides to take the server offline, you can simply move on to the next provider. And because both implement the same specification, you do not need to change your code in any way. You only need to replace the URL of the server to use.

NodeMCU

NodeMCU is definitely the cheapest development board introduced here that can be connected to the internet. On AliExpress, one of the biggest Chinese shopping platforms, you can get a NodeMCU for just a few dollars:

NodeMCU v3

Based on the ESP8266, it runs on 3.3V and offers many ports and supports protocols such as SPI or I2C. In contrast to the boards from Arduino or Particle, it only supports one analog input, though. If you plan on reading multiple analog sensors (for example, light and pressure sensors) at the same time, you should either get another board or you have to add an **analog-to-digital converter** (**ADC**) to your setup.

Be aware that there are various versions of the board to be found that are hard to differentiate from one another, one of them being the integrated USB-to-serial chip. Some boards use the CP2102, and others use the CH340 chip. Depending on which operating system you use, additional third-party drivers might be needed to get these chips working. You may also notice different version numbers, currently ranging from V1 to V3. The V1 board is outdated and you will probably not find it in shops anymore. V2 (Amica) had some improvements over V1, for example, a better form factor. And V3, also known as LoLin, does not have any major improvements but a big disadvantage—the board is considerably bigger than V2. If you place it on a breadboard, it will take up all of the available space in the narrower dimension, leaving no free rows on each side, so you cannot connect anything to its pins. A hacky workaround here is to place two breadboards next to each other and place the NodeMCU V3 board with one row of legs on the first breadboard and the other on the second. It's not good! Whoever decided to make these tweaks from V2 to V3 didn't put much thought into it.

While initially created to be programmed with the Lua programming language, you can use it in the Arduino IDE by installing the appropriate board in the Arduino board manager and programming it in the same way as you would with an Arduino using C++. If you need a super-cheap development board with slightly fewer options compared to an Arduino MKR 1010, you don't mind installing third-party drivers, and you don't plan on reading multiple analog sensors at the same time, the NodeMCU V2 Amica might be an option for you.

Raspberry Pi 4 Model B+

In contrast to the aforementioned development boards, which use a microcontroller, Raspberry Pi is a full-blown, single-board computer with separate memory, GPU (graphics card), and multi-core CPU in a very small form factor. It typically runs a Linux operating system, often Raspbian, a special version of Linux made for the Raspberry Pi. It was the first popular single-board computer combining both worlds—microcontroller boards with GPIO pins and a full operating system with support for USB devices such as hard drives, webcams, keyboards, or mice. It also ships with an Ethernet network port to create a stable network connection, an SD-card slot, and an audio jack connector, as well as an HDMI port to connect a screen to. It really is a tiny computer.

By now, there are a multitude of **system on a chip** (**SOC**) computers, but the Raspberry Pi remains the most important for IoT prototyping, because it has a huge community, so you will find a lot of hardware add-ons (called shields) and tutorials. You can see the predecessor to the Raspberry Pi 4 Model B, the Raspberry Pi 3 Model B+, as follows:

Raspberry Pi 3 Model B+

All of this might make it sound tempting to use the Raspberry Pi for every project, as it is so much more powerful than the others. But this comes at a cost. If you are just getting started with electronics and do not feel completely comfortable with Linux, there is a steeper learning curve compared to the microcontroller alternatives. You will spend time using the terminal instead of tools with a GUI, you'll fix issues here and there, and you'll code on the device itself or find a way of transferring your code from another computer to the Raspberry Pi (mostly via SSH, a network protocol). You also have to run some shell commands to make sure your code is executed when the Raspberry Pi is restarted—something that you don't have to worry about with less powerful alternatives such as Arduino, NodeMCU, or Particle Argon.

One thing to look out for, as with the NodeMCU development board, is the lack of analog ports—if you plan on using analog sensors, you need to either get an additional ADC or decide on another development board.

In general, if your project just uses the GPIO pins and does not require the extra power, the Raspberry Pi isn't the best choice. Use it whenever you need a camera, a screen, or one of the other options the Raspberry Pi offers over other boards.

In early 2019, some alternatives appeared that offer features that are similar to those on a Raspberry Pi board:

- **Coral Dev board**: The Coral Dev board (`https://coral.withgoogle.com/products/dev-board`) is a development board by Google designed especially for machine learning projects. It features a powerful Edge TPU coprocessor, which greatly accelerates machine learning tasks.
- **Jetson nano development kit**: The Jetson nano development kit (`https://developer.nvidia.com/embedded/buy/jetson-nano-devkit`) is a development board from the graphics card manufacturer Nvidia. Similar to the Coral Dev Board, it is intended for machine learning prototyping.

Arduino MKR WiFi 1010

The Arduino MKR WiFi 1010 is the successor to the Arduino MKR 1000 WiFi, which combined the functionality of the Arduino Zero with a Wi-Fi shield, and therefore lowered the entry barrier to creating internet-connected projects. Being an official Arduino product, getting started with it is really easy. You have to install the board, load the example sketch, and enter your router username and password, and your device is online. If you have ever worked with an Arduino Wi-Fi shield before, you probably do not have the best of memories as there were many things that could go wrong, leading to hours of debugging work. With the MKR series, this all became easier. The MKR 1000 also includes a battery socket, which makes building battery-powered projects even easier:

Arduino MKR WiFi 1010

The same is true for the Arduino MKR WiFi 1010. It is really easy to set up, has great library support, and brings some extra features. Using the newly added ESP32 chip, the MKR board is now also able to communicate via Bluetooth LE. Charging the battery on the MKR 1010 is done by connecting it to a power source via USB; it will automatically detect this and start charging the battery.

During this whole book, we will work with the Arduino MKR WiFi 1010. This will allow you to focus more on building and less on fixing problems on the side—using the Arduino IDE is currently the easiest way to get started with IoT prototyping, and it has great support for third-party libraries via its integrated library manager.

M5Stack

In some cases, all of the aforementioned development boards are overkill and what you really want is a boxed kit with some buttons and a screen. Various companies sell internet-connected buttons, so-called smart buttons, which have only one function: when you press the button, the device sends a request to a web server and therefore something happens. Use cases are versatile—with a press, you might order a product on Amazon directly (`https://www.amazon.com/b?ie=UTF8node=17729534011`), without the need to go through the whole process in a web shop.

Other smart buttons might be used for the elderly as an emergency button—pressing one will call an ambulance to an address that was set up previously (`https://flic.io/`).

M5Stack: a development board that can be used out of the box without the need for a breadboard or soldering

Another use case for a smart button might be to order your favorite pizza without the need to leave your sofa (other than to open the door when it is delivered). As you can see, there are many things that can be done just using one button with internet access.

The M5Stack, which is based on the ESP32 chip, includes not one but three buttons, which can be freely assigned: a screen; Bluetooth; Wi-Fi; an SD-card slot; an integrated battery; and a speaker, all neatly packed in its own sturdy case. Programming is done on the Arduino IDE. For this, you need to install the M5Stack library from its developers. It's the same process as setting up any other third-party development board in the Arduino IDE.

While the board looks like a closed unit, it is still hackable—using various connectors, you can make use of any of the available ESP32 input/output pins. You can also buy additional modules, which can be stacked onto the M5Stack to add functionality. Available modules include GSM—to add connectivity without Wi-Fi or GPS—so your device knows where it is located; battery extensions; and various sensors. The principle here is the same as with aforementioned module systems such as Grove, by Seeed Studio (`http://wiki.seeedstudio.com/Grove_System/`), or Qwiic, by SparkFun (`https://www.sparkfun.com/qwiic`): you create prototypes just by connecting plugs to sockets and writing code. No breadboard is required and no soldering.

So, if your next project can be done with components available to the M5Stack, it might be a really good option for a first prototype.

Summary

In this chapter, you were introduced to a lot of different principles, protocols, and microcontroller boards used in IoT prototyping. We've learned what constitutes the client-server architecture. Next, we looked at how the frontend and backend play together and how machines communicate with one another using APIs. Finally, we learned about the frameworks that use web development technologies such as HTML, CSS, and JavaScript to build desktop and mobile apps, which can communicate with your prototype. We also had a look at various communication methods; devices can transmit information not only over the internet, but also by using radio signals.

One of the most common modules for integrating radio-communication into your project is called XBee, and uses the ZigBee specification. Some devices also ship with Thread support—an upcoming radio-based protocol supported in recent microcontroller boards by Particle.

The Raspberry Pi is a good option if you want to integrate a screen or a webcam into your project, and NodeMCU is the most budget-friendly development board. Over the course of the book, we will use the Arduino MKR WiFi 1010 development board because it is very easy to work with.

We also had a look at operating voltages (3.3V and 5V) and found that not every input and output module is compatible with every development board.

I hope you don't feel overwhelmed by all of the development boards, principles, and protocols introduced in this chapter. You don't need to completely understand each and every one of them. This chapter should mainly give you an idea about the landscape of IoT prototyping and give you some pointers to explore further later on.

In the rest of this book, we will focus on working with the Arduino MKR WiFi 1010 as a development board and MQTT as a communication protocol; you will learn more about these in Chapter 3, *Getting Started with MQTT*.

Questions

1. Name three development boards suitable for IoT prototyping.
2. Does it matter if a development board runs on 3.3V or 5V?
3. Name four protocols used in IoT development.
4. What are over-the-air updates?

Further reading

- **Arduino MKR WiFi 1010:** https://store.arduino.cc/arduino-mkr-wifi-1010
- **Asus Tinker:** https://www.asus.com/de/Motherboards/Tinker-Board/
- **Amazon dash:** https://www.amazon.com/Dash-Buttons/b?ie=UTF8node= 10667898011
- **Cordova:** https://cordova.apache.org/
- **CRUD explanation:** https://www.codecademy.com/articles/what-is-crud
- **Electron:** https://electronjs.org/
- **Flutter:** https://flutter.io/
- **Smart button:** https://flic.io/
- **Grove:** http://wiki.seeedstudio.com/Grove_System/
- **If-This-Than-That (IFTTT):** https://ifttt.com/
- **IFTTT Webhooks:** https://ifttt.com/maker_webhooks
- **Ionic:** https://ionicframework.com/
- **MQTT versus CoAP video:** https://www.youtube.com/watch?v=pfG8uEDZj5g

- **NodeMCU version overview**: https://frightanic.com/iot/comparison-of-esp8266-nodemcu-development-boards/#v2
- **OpenThread**: https://openthread.io/
- **Particle**: https://www.particle.io/
- **Particle device cloud**: https://www.particle.io/device-cloud/
- **Particle mesh**: https://www.particle.io/mesh/
- **PhoneGap**: https://phonegap.com/
- **Raspberry Pi**: https://www.raspberrypi.org/
- **React Native**: https://facebook.github.io/react-native/
- **Sensors on sparkfun**: https://www.sparkfun.com/categories/23
- **Shiftr.io**: https://shiftr.io/
- **Sparkfun quiic**: https://www.sparkfun.com/qwiic
- **ThingSpeak**: https://thingspeak.com/
- **Thread protocol**: https://www.threadgroup.org/
- **ZigBee**: https://learn.sparkfun.com/tutorials/connectivity-of-the-internet-of-things/zigbee

Getting Started with MQTT

3

In the last chapter, you got an overview of the basic building blocks of IoT projects. We discussed the development boards that can be used for IoT prototyping, had a look at some existing web services that pair well with IoT prototypes, and gave an overview of protocols for IoT communication. One of these communication protocols is **MQ Telemetry Transport** (**MQTT**), which we will be using throughout this whole book, and which you will learn more about in this chapter.

You will learn about MQTT messages, how the underlying publish and subscribe pattern works, as well as special MQTT features that will enable you to use MQTT for a lot of different use cases. We will also have a look at the most common MQTT apps for iOS, Android, macOS, and Windows. Using one of these apps, you can easily interact with your IoT prototypes and exchange messages, for example, in order to read sensor data, or remote control your IoT devices.

The following topics will be covered in this chapter:

- Introducing MQTT
- Understanding the principles of publish and subscribe
- Exploring MQTT features
- Analyzing the security of MQTT servers
- MQTT servers and cloud providers
- Comparing MQTT iOS and Android apps
- Exploring MQTT desktop apps
- Understanding MQTT libraries

Introducing MQTT

MQTT is a protocol for machine-to-machine communication that was invented in 1999 by Dr. Andy Stanford-Clark (IBM). Because of its lightweight nature, it is especially useful when used with microcontrollers for IoT projects. It is lightweight in terms of both energy consumption and bandwidth because it does not have much overhead, which makes it a great fit for battery-powered projects on microcontroller boards.

MQTT is used in a wide range of projects, from small-scale DIY home automation to applications in the healthcare sector, where practitioners use it to communicate with medical devices (for example, blood pressure monitors); and by oil companies to monitor miles of pipelines. Even Facebook uses it for its Messenger app.

In 2014, MQTT 3.1.1 was officially accepted as an OASIS standard, which sent out a great signal to everyone who was already using or considering using MQTT, implying that it is here to stay as an open standard, and that it won't go away easily.

MQTT, by default, runs on two ports, which were assigned by the Internet Assigned Numbers Authority: the `1883` port for MQTT, and the `8883` port for secure MQTT.

The version number, 3.1.1, specifies the version of the specification. Projects that implement the specification, for example, for an Arduino library, have to follow this specification. You will get to know more about the various versions of MQTT, and what you need to look out for when using such a library later in the chapter.

The essence of MQTT is its publish/subscribe pattern. Think of a newspaper here: when opening the newspaper, chances are high that you skim through the pages to get to a certain category in the newspaper. Every person has different interests. Let's say your interests are science and the economy. You don't care about the rest of the articles to be found in a newspaper.

Sometimes, it might happen that, after you find the appropriate sections in the newspaper, there is no content. The section simply does not exist in this edition because, on this particular day, there is nothing new to report about. Maybe tomorrow, or maybe the day after, you will be lucky enough to satisfy your thirst for knowledge.

A technology called **RSS** (which means **Rich Site Summary**, sometimes called **Really Simple Syndication**) brought this to the web in the 90s. When a website support RSS, you can add it to your feed-reader—an application to keep track of all of the websites (in terms of RSS-called feeds) that you are interested in; whenever there is a new article in the area you are interested in, it will pop up. There's no need to browse through tons of articles that you don't care about, and no need to navigate to a website manually, just to find out that there is no new content; you simply **subscribe** to a topic.

MQTT does exactly this. Each smart device can subscribe to the content that it is interested in, or that it needs in order to operate, accordingly. To subscribe to a topic, you need to know the topic name (for example, `/science`), but more on that later. Additionally, each device can **publish** information on its own, which others can then subscribe to. Each device is both a sender and a receiver.

Before we dig any deeper, let's first have a look at what traditional client-server communication would look like without MQTT. Let's say we want to build a lamp with a physically separate on/off switch using two Arduinos. Physically separate means it can be taken with you, and used from another city, country, or continent.

Arduino A—the controller—has a physical button; Arduino B—the lamp—has an LED. When the button is pressed, the LED on the other Arduino should switch from on to off, or the other way around. In its simplest form, this would mean that the controller sends `1` when the lamp on the other Arduino should go on, and `0` when it should go off. Because the controller is communicating directly with the lamp, it needs to know how to reach the lamp, or where to send that `1` or `0` to in order to control it. This can be done using its IP address (for example, `203.0.113.193`) and port (for example, `3000`). So, somewhere in the code of the controller, this IP address needs to be hardcoded. If the lamp does not have a static IP address (which is very likely), the lamp would stop working when the internet service provider assigns a new IP address, which, in Germany, typically happens every 24 hours. Also, the IP address will change when the controller is connected to another network.

In addition to the problems with dynamic IP addresses, if we want to add a second, third, and fourth lamp to our setup, we need to edit the code of our controller each time. At first, this code might look something like the following:

```
if (magicButton == 1) {
  // send command to turn lamp on
}
```

After adding our fourth lamp, it might look like this:

```
if (magicButton == 1) {
  // send command to turn first lamp on
  // send command to turn second lamp on
  // send command to turn third lamp on
  // send command to turn fourth lamp on
}
```

Every time, we would need to edit the code of the controller. Wouldn't it be nice if we could just flash the lamp code to another Arduino in order to integrate a fifth lamp into our setup, without making any modifications?

MQTT makes these tasks so much easier! The controller publishes an on or off message to the /magic button topic, and all of the lamps just subscribe to that topic (more on topic names later). Every time the button is pressed, the lamps will get an update and turn on or off, accordingly.

Another problem that might occur with every internet-connected device is that the network connection gets disturbed, which might lead to one lamp not getting the message to turn off. Network packets sometimes just get lost. The lamp would stay on until the button is pressed twice (one time to send the on command, which was already sent, and another time to send the off command, to turn off all lamps):

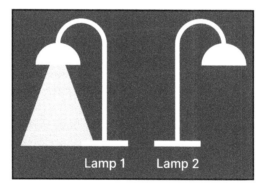

One lamp is still on, because it did not receive the command to turn off

While there are certain ways and other technologies that can be used to overcome some of these problems (for example, DynDNS, which is out of the scope of this book, or CoAP, which we had a brief look at in Chapter 2, *Basic Architecture of an IoT Prototype*), MQTT definitely makes this the easiest.

At the center of each IoT application is the MQTT server, which all of the MQTT clients connect to. It acts as a hub for the message exchange, taking in published messages and subscriptions. Every time it receives a new message, it checks whether one of the MQTT clients is interested in the published message (if it is subscribed to this specific topic) and forwards the message, accordingly.

Let's look at the aforementioned problems one by one.

Problem 1 – dynamic IP addresses

The dynamic IP addresses of the lamps lead to code changes being necessary in the Arduino controller. Using MQTT, the controller does not talk directly to the lamps. It does not know anything about them. Instead, it just publishes on or off to the /magic button topic of the MQTT server (the central entity, which manages all published messages and subscriptions), without knowing whether there are any lamps listening—basically saying *to whoever it may concern, the magic button was pressed and it is on now*.

All that the controller Arduino and the lamps need to know about is where to find the MQTT server (typically, using a third-party server with a static IP), as well as the topic where the information is published to, in this case, /magic button.

Problem 2 – code changes

MQTT makes code changes easier to scale. When using MQTT, there is always a centralized MQTT server, which keeps track of all of the clients, messages, and subscriptions. To add another device to our network, we don't have to modify the code of the other devices in order to receive information from them. It is best to understand this principle by thinking about communication between humans. If you have some information that you want to pass on to multiple people, there are two ways to let them know about it. The first way, which can be compared to traditional infrastructure on the web, is one-to-one communication. You let each of them know individually about what you have to say. The other way is letting them know all at once, basically screaming it (in a non-aggressive, creepy way). If somebody else is close by while you scream it, they will also hear it. Just because somebody else hears it too, does not make it any more effort for you. You say it once, and whoever is interested in the information takes notice.

Adding a new lamp to your MQTT network can be compared to that. It does not make a big difference; it just works. No code changes from the other devices on the network are needed in order to integrate it. The new device specifies which information it is interested in, and the MQTT server makes sure that the new device also gets notified when there are new messages on one of the subscribed channels.

We need to make sure that each client has a unique client ID, so that the MQTT server can distinguish between them. For this, a minor code change on the new client is needed when connecting to the MQTT server.

Problem 3 – network disturbances leading to lost messages

MQTT is shipped with a feature called **Quality of Service** (**QoS**), which can be used to ensure messages are delivered correctly. When a message is lost in transit, it will be re-sent until it is delivered successfully.

Let's stick with the example of the smart lamp for a bit—a lamp that can be controlled by another device over the internet with a simple button, which turns it either on or off when it's pressed.

Imagine you go on vacation. In your bag, you take the smart lamp with you. While on the airplane, there is no internet connection (sadly); therefore, the lamp is offline, and does not get any updates from the controller device. While you are on the airplane, the controller device sends a message to turn the lamp on. Because it is currently offline, this information gets lost.

After the plane lands, you arrive at the hotel, plug the lamp into a power outlet, and connect it to the local network. It will stay off, because the message to turn it on was sent while the lamp was offline. Not a big problem, you might think. But bear with me, this is just an example, and relates to more complex scenarios as well, where the internet connection is lost for some time.

This situation, where the lamp is in this weird state of thinking it is off, while the controller thinks it is on, leads to problems. To turn it on now, the button on the controller has to be pressed twice, instead of once. One time to send the off command (which the lamp will ignore because it is already off), and another time to send the on command to really turn it on.

Wouldn't it be nice if there was a buffer, and the lamp would automagically receive the command again, once it is back online? Using MQTT and a QoS setting of 1 or 2, this is possible. Messages will be retained until they are delivered correctly.

> MQTT servers have different limits for how long, or how many messages, can be kept for each client. Also note that not every MQTT server supports message buffering for when a device goes offline. It produces more overheads, because the server needs to keep track of every message.

Understanding the principle of publish and subscribe

The principle of publish and subscribe is the heart of MQTT. Devices can publish messages (for example, `on`) to a channel that can be freely defined; for example, `/living-room/my-custom-coffee-machine`.

Other devices that are connected to the same MQTT server can subscribe to this channel, resulting in near instant updates, whenever there is a new message published to that particular channel.

The MQTT server connects all of the publishers and subscribers, and keeps track of the subscriptions for each client. Every client only receives what it is subscribed to:

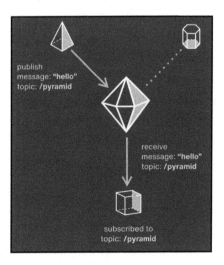

Three devices connected to an MQTT server

All devices are connected to the MQTT server (in the middle). It forwards the published messages to the subscribers.

Sometimes, publish and subscribe is referred to as pub/sub. One of the earliest MQTT clients for Arduino is called pubsubclient.

At the most basic level, in order to publish a message, two things need to be given: a topic name (where to publish stuff to), and a message (what to publish). The message can also be empty. There are a few more things to consider when you publish a message (for example, QoS), which we will have a closer look at later.

Topic names

The topic name specifies where MQTT messages are sent to. Each MQTT server has its own topics, which can be freely defined on the fly by the publishers. Imagine a topic as a tree of categories. Each / defines a sub-category. So, when we look back at the topic name from the previous section, /living-room/my-custom-coffee-machine, we can see that the topic consists of two components:

- /living room
- /my-custom-coffee-machine

If, next to your custom smart coffee machine, you had another smart device in the living room, which communicates via MQTT, let's say a custom smart light, you could publish its information in the same /living-room root topic: /living-room/my-custom-smart-light.

Topic names are case-sensitive. So, /livingroom, /Livingroom, and /LIVINGROOM are three different topics.

As with URLs on a website, the / character is used to separate the different parts of a topic. There are very few restrictions about what a valid topic name is. It needs to be a UTF-8 string, which can include spaces or special characters. There are a few exceptions though. Three characters have a special meaning in MQTT:

- /: Separator for the specific parts of your topic name
- #: Multi-level wildcard
- +: Single-level wildcard

> Wildcards can be used to subscribe to various channels at once, but more about that later.

Apart from these three characters, you are free to use whatever you want. A note of caution though: just because it is allowed, does not mean that it is a good idea to use the full range of UTF-8 characters. Some MQTT libraries and servers that are not fully compliant to the specification might not support the full range of characters. Also, using spaces can result in some problems in other cases. If you want to play safe, stick to these characters: a-Z, A-Z, 0-9, _, and -. Feel free to use the full range though. If you want your topic names to be emojis, go for it. Any of the UTF-8 emojis are a valid character, and therefore can be used as a topic name. For a list of UTF-8 emojis, see `https://unicode.org/emoji/charts/full-emoji-list.html`. When pasting any of them in the Arduino IDE or another code editor, they will probably be rendered incorrectly (as a square), but should be rendered correctly in a web browser or app where you could subscribe to the topic.

As you can see, the following are some valid topic names:

```
/
/kitchen
/kitchen/smart-coffee-machine
/kitchen/smart coffee machine
/Kitchen/Smart Coffee Machine/Temperature in °C
/kitchen/shelf/middle-drawer/my-hidden-smart-thing
```

Next, let's look at wildcards.

Wildcards

When publishing an MQTT message, you always need to specify the exact topic where it should go, as in the following example:

```
/kitchen/smart-coffee-machine/temperature
```

Let's say that the smart coffee machine with MQTT has various channels that can be used to control it and receive its status (for example, the current temperature):

```
/kitchen/smart-coffee-machine/temperature
/kitchen/smart-coffee-machine/status
/kitchen/smart-coffee-machine/make-coffee
```

If we had another smart device to monitor our ceiling light and smart coffee machine, we could then subscribe to multiple channels:

```
/kitchen/smart-coffee-machine/temperature
/kitchen/smart-coffee-machine/status
/kitchen/ceiling-light/status
```

If, one year after you finish reading this book, you build yourself an army of smart home devices that you want to control via MQTT, this list would grow further. Luckily, MQTT allows us to use wildcards while subscribing to topics, which allows us to subscribe to multiple topics at once. There are two different wildcard characters that you can use in your topic filters.

Multi-level wildcards

Multi-level wildcards, indicated by a number/hashtag character (#), can be used to subscribe to various topics at once—as the name suggests—multiple levels deep. Below, you can see individual topic names that we could subscribe to one by one:

```
/kitchen/smart-coffee-machine/temperature
/kitchen/smart-coffee-machine/status
/kitchen/ceiling-light/status
```

But, instead of subscribing to them individually, we could simply subscribe to them all at once using a wildcard character:

```
/kitchen/#
```

This means *I want to subscribe to everything that happens in this channel and all of its sub-channels*. As the name suggests, this is multiple levels deep, so all of these messages would be included in this topic filter:

```
/kitchen
/kitchen/smart-coffee-machine
/kitchen/smart-coffee-machine/temperature
/kitchen/smart-coffee-machine/status
/kitchen/smart-coffee-machine/make-coffee
/kitchen/ceiling-light
/kitchen/ceiling-light/status
/kitchen/ceiling-light/turn-on
/kitchen/ceiling-light/turn-off
```

If we later decide to add another device that publishes to the `/kitchen/my-new-smart-device` topic, we would automatically receive its messages, as well.

To only subscribe to all messages published by the smart coffee maker, you could subscribe to `/kitchen/smart-coffee-machine/#`.

`#` must always be the last character in the topic filter, so you cannot use a filter like this: `/kitchen/#/status`.

Also, it must replace a topic section completely, so `/kitchen#` would not work, while `/kitchen/#` would.

Single-level wildcards

Similar to the multi-level wildcard, the single-level wildcard, indicated by a plus sign (+), can be used to subscribe to various topics at once, but only one level deep. In contrast to the multi-level wildcard, it can be used at any level of the topic name—at the beginning, middle, or end. It can even be used multiple times per topic filter, as well as being combined with the multi-level wildcard. All of the following uses of single-level wildcards are valid:

- `/+`: This subscribes to the following:
  - `/kitchen`
- `/kitchen/+/status`: This subscribes to the following:
  - `/kitchen/smart-coffee-machine/status`
  - `/kitchen/ceiling-light/status`

- `/+/ceiling-light/+`: This subscribes to the following:
  - `/kitchen/ceiling-light/status`
  - `/kitchen/ceiling-light/turn-on`
  - `/kitchen/ceiling-light/turn-off`
- `/+/ceiling-light/#`: This subscribes to the following
  - `/kitchen/ceiling-light/status`
  - `/kitchen/ceiling-light/turn-on`
  - `/kitchen/ceiling-light/turn-off`

Similar to the multi-level wildcard, the single-level wildcard must replace a complete topic segment and cannot be combined with one. So, `/kitchen+/status` would not work, while `/kitchen/+/status` would.

Forbidden characters

There are a few exceptions to characters that cannot be used in topic names: topic names beginning with a dollar sign (`$`) are treated differently, and are often used on MQTT servers to transmit internal server messages (using the `$SYS` topic). You should not use the `$` sign as the first character of a topic name in your projects.

Also, you should not use the null character in topic names (`0x00`). If you don't know what it is, don't worry. It is a control character that is used in languages such as C, where it terminates a string. It is not a physical key on your keyboard.

Exploring MQTT features

In this section, we will have a closer look at some important features of MQTT:

- **QoS**: For message buffering (when offline).
- **Last will/testament**: To let the network know when a device goes offline; for example, due to empty batteries.
- **Keep alive**: To define how often each device needs to call back home to let the MQTT server know that it is still online.
- **Persistent sessions**: To store various information on the MQTT server, while the client is offline.
- **Retained messages**: To keep messages available for new subscribers.

Quality of Service

MQTT supports a feature called **QoS**, which makes it possible for clients to receive messages that were sent while the device was offline. You can see this feature being used in messenger apps (for example, Facebook Messenger, which also uses MQTT under the hood). When somebody sends you a message while you are offline, you still receive the message later on, when you come back online; it does not get lost.

There are three different settings for QoS:

- **QoS 0**: Fire and forget
- **QoS 1**: Deliver at least once
- **QoS 2**: Deliver exactly once

When publishing messages, as well as when subscribing to messages, each client can independently decide which QoS to use. So, simply publishing with QoS 1 or QoS 2 does not make sure each subscriber really gets the message; it only makes sure the server receives the message. From there on, each subscriber is responsible for their own, and decides how important the messages are to them. This is important to keep in mind: messages are published from the publisher to the server using a QoS that the publisher decides on, and are later on forwarded from the server to the subscriber using a QoS that the subscriber defines.

QoS 0 – fire and forget

Messages sent with QoS 0 are officially labeled as message delivery: they, at most once, are sent without caring if they ever reach the subscriber(s). It is the equivalent of sending a letter by post—it probably will be delivered, but something could go wrong. You never know for sure if it fell off a box and went missing.

While *at most once* does not sound very reliable (I would not send my packages with a postal service that claims they deliver their packets at most once), in most cases, MQTT packets sent with QoS 0 are delivered correctly anyway.

If you are wondering about the *at most* part, it just means that messages won't be delivered twice, or even more times, which can happen under certain circumstances when using QoS 1 (at least once). While in a perfect world, packages would never get lost and would always be delivered exactly once, there is always a trade-off between performance and reliability. QoS 0 is easier to handle for MQTT servers than QoS 1 or 2, because the server does not have to keep track of whether messages are delivered correctly or not:

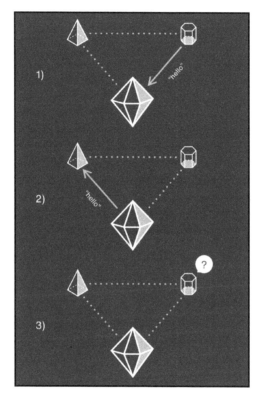

QoS 0: sometimes messages might get lost

When using QoS 0, the sender does not know whether the message has been delivered correctly, or not. In the preceding diagram, the big shape represents the MQTT server; the small ones are clients. Clients communicate via the server, not directly with each other.

QoS 1 – at least once

QoS 1 is more reliable than QoS 0. When a device goes offline (because of a power outlet, empty batteries, or any other reason), a message sent to it while it was offline will be stored in a message buffer on the MQTT server. When the device goes back online, the MQTT server will deliver all of the missed messages. Until the device comes back online, the server will keep hold of them.

Sometimes, messages are delivered more than once when using QoS 1. Here, it really depends on whether this produces any problems in your use case. Coming back to our smart lamp example, which can be either turned on by sending it an on command or off by sending it an off command, sending either of these commands twice is no problem.

QoS 1 and QoS 2 both use message buffering on the server, so they store the messages until the client comes back online. The only difference is in the way the acknowledgment works. QoS 1 uses fewer resources than QoS 2, with the downside that identical messages can be delivered multiple times by accident.

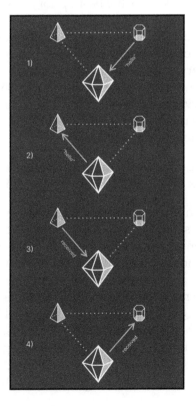

QoS 1: messages are re-sent until they are delivered

When using QoS 1, messages will be delivered for sure. Sometimes, by accident, they will even be delivered multiple times. In the preceding diagram, the big shape represents the MQTT server; the small ones are the clients.

QoS 2 – exactly once

If dealing with duplicate messages is not an option, and you want to make sure that you receive messages on the client, even if it was offline for some time, then QoS 2 can be used. The downside of this is that it uses more resources than QoS 0 and QoS 1, and should, therefore, be used with care. Under the hood, more messages need to be sent to make sure published messages are delivered only once and that the subscriber receives the message, even if it was offline for some time.

How to choose the best QoS

You might think now that using QoS 2 is the best option, so you always know for sure that every message is delivered correctly. It's not that easy. You can put QoS 0, 1, and 2 on a scale. On the left-hand side of the scale is maximum performance; on the right-hand side, maximum reliability. QoS 0 is on the outer far-left side, being the fastest of the three, and the easiest to handle for the server. QoS 2 is the other extreme on the right-hand side. It is the most reliable option. QoS 1 is somewhere in the middle. While it supports offline delivery, and therefore the server needs to keep track of whether each message has been delivered correctly, it is not as reliable as QoS 2. QoS 1 can lead to messages being delivered more than once, instead of exactly once.

Under the hood, a lot more messages need to be sent between publishers and subscribers to assure that every message was delivered correctly when using QoS 1 or 2.

The good thing is, you don't have to decide now which QoS you are about to pick for every message that you are about to send. For each message that you publish, you can decide which QoS makes the most sense. Ask yourself: *does it matter if this message is skipped?* If the answer is "no," you should choose QoS 0. A great use case for QoS 0 is when you send out sensor data each second. Here, it is mostly fine when one message gets lost; it does not matter too much. If you really need to make sure a message is delivered correctly, you have to choose between QoS 1 and QoS 2. Because QoS 2 means a bit more extra work over QoS 1, you should choose QoS 1 when it does not matter if a message gets sent multiple times. Imagine you want to control a light bulb via MQTT—sending it on two times does not do any harm—it is already on, and it will simply stay on. If, on the other hand, messages that get delivered twice or more produce problems, you should choose QoS 2, but this should be the exception.

Some MQTT servers only support QoS 0, or QoS 0 and QoS 1. If your project relies on a specific QoS setting, you need to check whether the MQTT server and the library that you are using support it.

Last will messages

Network connections can be volatile, and there are many things that can lead to them being disconnected—temporarily or permanently. Not only can a connection fail, but a device might run out of power (blackout or empty batteries). MQTT has an option called **last will/testament**, which makes it possible to publish a message when a device goes offline: a protocol or I/O error occurs when the device does not call back in a certain amount of time (the keep-alive time), or the application gets terminated without being disconnected from the server properly.

Similar to a regular MQTT message, a last will message is sent to a certain channel. A very good use case for the last will feature is posting an update to a status channel. Every time the device goes offline (for any of the reasons specified previously), other devices are notified:

"To whom it may concern, I am not online anymore. Please don't rely on me working. Once I am back online, I will let you know."

The topic name, /status, can be freely defined (as with all topic names). Messages in a /status topic could look like this:

```
TOPIC                        MESSAGE
/my-smart-device/status      connected
/my-smart-device/status      disconnected
```

Other devices can subscribe to this channel, in the same way as they would subscribe to another channel, and this way, stay up to date if another device is online or offline.

Keep alive

The MQTT server and client need to know whether each other is still online at all times; for example, for features such as *last will* to work. Every time a client publishes a message or subscribes to a topic, the server gets a life signal from the client. But what if the client does not send something for a while, or subscribes to a new topic? Under the hood, implemented by the MQTT library and server, ping packets are sent to let each other know that the connection is still stable and both are still online, ready to exchange messages.

If the server does not receive a life signal from a client, either via a normal publish/subscribe call or an empty ping message, the server will disconnect the client. When this happens, and the client is disconnected by the server (because it did not call home in the **keep-alive** interval), it has to reconnect. To fine-tune this behavior, the keep-alive setting exists, which can be manually set when connecting to an MQTT server. Typically, this value should be set to a few seconds, but in most cases, it will be fine to not set it at all, and to rely on the default value, which depends on the implementation of the library used.

Allowed values are between a few milliseconds to roughly 18 hours, which is way too long for most use cases, and should be avoided. When using 0 as a keep-alive value, the feature is turned off completely:

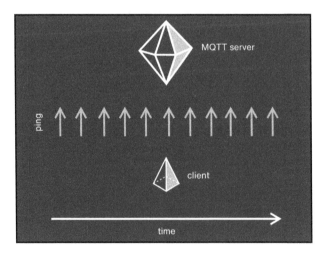

The client has to ping the server at certain intervals

Next, let's have a look at an important backbone of MQTT—persistent sessions/clean sessions.

Persistent sessions/clean sessions

We have already discussed QoS, which makes it possible to receive messages, even when the recipient is offline. When a device is unavailable, the MQTT server will keep track of it, and re-send the message once the device goes back online again.

To make this possible, the server needs to do some bookkeeping. For every message that was sent using this feature, the server needs to keep track of whether the message has already been delivered to the recipient or not. Once it has been delivered, the server will forget about it. Imagine a mail. If you are not at home when they try to deliver a package to you, they will try again another day. Once it has been delivered successfully, they will forget about it.

The server needs to store the following information for each client who is using this feature:

- Messages that have not been delivered yet and haven't been acknowledged by the subscribers (using QoS 1 or QoS 2)
- Messages that have not yet been sent to the subscribers (using QoS 0)
- All subscriptions by the client

When a client reconnects to a server, it does not need to resubscribe to any topics, as the subscribed topics are stored on the server, and therefore it will continue receiving new (and pending QoS 1 and QoS 2) messages on these topics.

For this to work, the client must identify itself in the same way, so the server says, "*Ah, it's Courtney the Coffee Machine again. Wait, let me check if I have some messages for you...Here you go! (Passing messages on to client.) Ah, I also noticed you are interested in messages in the /button-pressed topic; I will keep you updated.*" So, the client always needs to connect with the exact same client name; otherwise, the MQTT server thinks that it is a new device, without prior sessions. If you use a lot of MQTT devices at the same time, you also need to make sure that you don't use the same client name for multiple devices. This could have various negative effects, which depend on the implementation of the MQTT server. So, always use one client ID per device when connecting to an MQTT server.

An exception to persistent storage is using a clean session setting. When connecting to the MQTT server, you can request a clean session, basically saying, "*Forget everything you know about me, I want to start from scratch. I'm not interested in any prior messages or subscriptions.*" While developing a smart device using MQTT, you might end up with a lot of subscriptions due to rapid changes and iterations. To overcome this, you could reconnect once to the server with the same client ID and a clean session (set to `true`), which will delete all of the information stored about the client. You could then reconnect without a clean session.

So, which of the two should you use: a clean or persistent session? Well, it depends. If the client is only publishing messages, but does not depend on any incoming messages (no subscriptions), you probably do not need a persistent session, and should use a clean session. If the client, on the other hand, depends on incoming messages, which should be received even if the device went offline, or there was an error leading to a reconnection, then you might want to use a persistent session. Keep in mind that every use of persistent sessions and the usage of QoS 1 and QoS 2 messages (offline delivery), put additional work on the MQTT server, because it needs to keep track of every message and whether it has been delivered. There is also a limit as to how many messages the MQTT server can store for later delivery. Imagine you go on vacation and receive one letter per day. When you come back after 2 weeks, there might still be some space left in your mailbox, but if you go on a world trip, it will be bursting at the seams.

If you only use a couple of devices, and don't publish to topics with subscribers using offline delivery, which do not connect for a long time, everything is fine. Even if you do so, you will most probably use a third-party MQTT server (we don't need to do everything ourselves), and the server administrator will surely define rules about what should happen if the message buffer is exceeded.

Retained messages

If a client subscribes to a topic, let's say, `/kitchen/ceiling-light/status` (which is either on or off), using QoS 1 or QoS 2, it will get an update whenever the status changes from on to off, or the other way around, even after a reconnect. But this only works when there is an active subscription and QoS 1 or 2 are being used. Imagine that you integrate a new smart device into your setup—a Raspberry Pi-powered status monitor with a display to visualize all of your IoT devices together, so that you know which among them are online or offline. Because this new device does not have any subscriptions, there is no way for it to find out whether the ceiling light is currently on or off. Here, retained messages come into play. A retained message is like a regular message, but it sticks to a certain topic, such as a welcome message. Every time a new client subscribes to the topic, it will instantly receive this sticky message, even if there were no active subscriptions before, and no persistent session existed. There can only be one retained message per topic, so if you republish a retained message to the same topic, it will overwrite the last retained message. Deleting a retained message depends on the library used—in theory, publishing a zero-byte message to a topic deletes the retained message.

In contrast to buffered messages (QoS 1 and QoS 2), retained messages are friendly on the MQTT server. Because there can only be one retained message per topic, the message buffer will not be exceeded easily.

You can still be kicked out by the third-party MQTT server if you publish too many messages in a short amount of time. Make sure that you always wait a bit (for example, a couple of seconds) before you publish a new message. This is only relevant when sending out continuous data; for example, sensor values (where using QoS 1 and QoS 2 should also be avoided).

If a device only needs the last valid value, and not the values that were published before (while the device was offline), using retained messages is a good fit. If new devices also need to have access to the last valid value (when connecting for the first time), using retained messages is the only way.

Analyzing the security of MQTT servers

There are four major ways to communicate with MQTT servers: MQTT, secure MQTT, MQTT over WebSockets, and MQTT over secure WebSockets. WebSockets are typically used when websites communicate with an MQTT server on the frontend.

Let's talk about the secure element in secure MQTT and secure WebSockets.

This book is all about prototyping: trying things out and failing (or succeeding) fast. For this, we don't care too much about security, as properly securing our prototypes would increase complexity and require more time. But, we should understand how to establish a basic level of security, and know when basic security is not enough.

SSL/TSL

When using secure MQTT, or secure MQTT over WebSockets, we have to deal with certificates, which would take up a huge portion of this book on its own, leaving us with no time for prototyping. Knowing about possible security issues is important, but unless your project is used permanently, or you work on prototypes for clients that will be used in production, it is not worth the hassle of setting up and updating SSL certificates, which are a requirement for a secure connection. I don't want to advise you to be careless. You should not update your microwave to be controlled via MQTT or create a door lock that communicates over an unsecured MQTT connection. This would be extremely careless and could result in some serious damage.

Username and password

You will get a basic level of protection using username and password authentication when you connect to the MQTT server. In the hands-on projects, we will use shiftr.io as the MQTT server, which makes it possible to restrict access to a personal namespace to only the accounts that you granted access to. Compared to transport-level encryption with SSL/TSL, this is just basic encryption. The username and password are transmitted in an unencrypted state, from your client (in our case, the Arduino) to the MQTT server. Therefore, third parties could sniff the password, and use it to log in to the MQTT server themselves.

In this whole book, we will be using the MQTT shiftr.io server. It will be introduced in the *MQTT servers and cloud providers* section.

On shiftr.io, we will authenticate using a default username and password (shared by all users), which is fine for trying things out. But, you really should consider using your protected username and password afterward, and create a private namespace that only you have access to.

Message/payload encryption

When not using SSL/TSL, you could increase security by encrypting the messages (payload) that are sent from your client to the MQTT server. Everything you send when not using TSL/SSL can potentially be read by third parties. Obfuscating your messages might help you to avoid revealing the meaning of your messages. A naive obfuscation algorithm could be to change each character to the next one in the alphabet. Let's say you want to send a message: `Hello`. Changing each character to the next one in the alphabet would turn it into `Ifmmp`. Still not very secure, but hackers would need to at least invest a little bit of time in order to make sense of your messages.

 Using payload encryption in this way will only encrypt the payload (message), but not the topic. If your topic is `/smart-door/open` with the `true` message, message encryption will not help much, because the meaning of the message could be easily guessed.

Definitely a better, and more secure way, of achieving client-side encryption is using an encryption library. There are various Arduino libraries that can be used for client-side encryption; for example, Cape (`https://github.com/gioblu/Cape`).

Security recommendations

As long as you are prototyping, and no harm can be done when your device is being accessed by third parties, you should protect your messages via a username and a password, which offers basic security. For most prototyping needs, it is probably secure enough.

If you want to add another layer of security, you can encrypt the MQTT payload with an encryption library.

 Ask yourself: what could happen if somebody else gained access to my device?

If you are really serious about encryption (which you should be, if real damage could be done if your device was hacked), you should further research how to use SSL/TSL with Arduino MKR WiFi 1010.

MQTT servers and cloud providers

The center of each MQTT project is the MQTT server, sometimes referred to as the MQTT broker. It is the control center to which each device in the MQTT network needs to connect, where all of the MQTT messages are sent to, and where devices can subscribe to the topics that they are interested in. It handles the topic filtering, so it knows which devices are interested in which topics, and forwards the messages accordingly. It also keeps track of messages that could not be delivered yet; for example, because a subscribed device went offline. The server will try again, until the device goes back online and the message can be delivered.

Because MQTT is an open protocol, you can freely switch between MQTT servers. In contrast to proprietary solutions, where you cannot easily switch servers, MQTT makes this very easy. In most cases, you just have to change three lines of code in order to switch to another MQTT server:

- Username
- Password
- Server URL

There are many open and closed source implementations of MQTT servers to choose from. In the following sections, we will have a brief look at a few of them.

 There are also a number of free MQTT servers that you might use for prototyping, and they are available on the official MQTT GitHub page (`https://github.com/mqtt/mqtt.github.io/wiki/public_brokers`).

Additionally, we will have a look at the landscape of MQTT libraries, which are available for every major programming language.

Mosquitto

Mosquitto (`https://mosquitto.org`) is an open source MQTT 3.1.1 server, which is part of the Eclipse Foundation; you might know this from other open source tools, such as the Eclipse IDE, one of the most used Java IDEs.

Mosquitto offers various command-line tools that make it really easy to try out all of the features MQTT has to offer on your local machine. In the *Installing Mosquitto* section, which you can find in Chapter 4, *Setting Up a Lab Environment*, we will set up a local Mosquitto server, in order to get our hands dirty with MQTT.

AWS IoT/Google Cloud

Neither Amazon AWS (`https://aws.amazon.com`), nor Google Cloud (`https://cloud.google.com`), offer regular (unsecured) access to their MQTT servers, which makes them unsuitable for the purpose of this book. When your prototypes mature, and you start thinking about securing your MQTT connections, it is worth having another look. This is especially true if you need to think about scaling when you add more and more devices to your network (hundreds or thousands).

shiftr.io

shiftr.io (`https://shiftr.io`) is a very unique web service that is based around MQTT. It was developed by Joël Gähwiler at the Zurich University of Arts, in the Interaction Design faculty. Its main audience are not multi-million-dollar companies, which Amazon and Google might want to attract with their MQTT services, but artists, designers, makers, and developers, who want to evaluate a prototype in the most straightforward way:

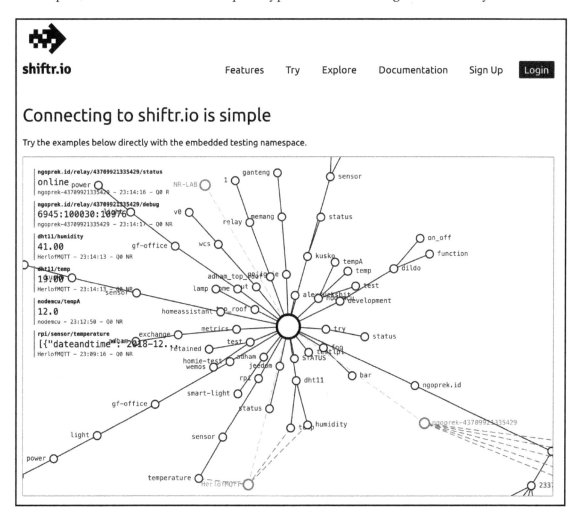

This is an image of shiftr.io, which offers a unique real-time visualization for inspecting MQTT traffic; the text and numbers in this image are intentionally illegible

shiftr.io offers various ways to interact with the underlying user accounts—via MQTT, secure MQTT, MQTT over WebSockets, and MQTT over secure WebSockets. Using MQTT over WebSockets, we can add communication via MQTT to a web frontend—so, the part of a website that runs in your browser.

What makes shiftr.io so unique is its simplicity. It offers all of the features that we will need in an uncluttered web interface, and visualizes the messages sent within our namespace. This makes debugging a breeze. By looking at the graph and its animated circles (each representing an MQTT topic segment), we can see whether the message flow is working as expected, or maybe one of our devices went silent (offline, or completely nuts—sending 1,000 messages per second, which will eventually lead to the client being kicked out of shiftr.io).

shiftr.io offers free accounts that can be used for public MQTT communication. Public, here, means that anybody could potentially read and write to the same channel, which, for a prototype, is probably fine. For serious applications, this is not an option, as it is a big security risk. shiftr.io also offers private namespaces, which only you and the devices that you authorize have access to. Currently, with a free account, you are also able to create private namespaces, so that you can keep your MQTT communication for yourself, but this might be a paid-only feature in the future. For early prototyping, using the public namespaces is fine. In this book, we will make extensive use of shiftr.io, as it is the easiest-to-use MQTT server, and it offers graphical debugging/flow visualization.

Comparing MQTT iOS and Android apps

The beauty of MQTT is—apart from the things that you already know about—its landscape of servers, libraries, and apps with MQTT support, which you can mix and match in any way that you like.

It has never been this easy to control your Arduino project from your smartphone, without the need to write additional iOS or Android code. For this, you can utilize one of the many iOS and Android apps that connect to your MQTT server of choice, in the same way that Arduino does. It's just another MQTT client.

Apps with MQTT support serve two main purposes:

- Logging/visualization/inspection of your MQTT devices and their data (passive).
- Controlling your MQTT devices and sending commands (active).

While some apps exist that work purely using text (MQTT topic and message), and that are especially useful for debugging, there are a few that can be used to create custom user interfaces to control your IoT devices, or to visualize their data on unique dashboards.

We will now have a look at a few of those apps, which are freely available, and which you can use to control your MQTT-powered projects from wherever you are, or to check their status when not at home (for example, if all of your devices are still online). These apps should give you a glimpse of how easily you can enhance your prototypes with custom control surfaces, either for demonstration purposes, or just for convenience. Going one step further, a custom website, or app, could be built that uses an MQTT JavaScript library to communicate with your MQTT-powered prototypes.

MQTT Dash

MQTT Dash (`https://play.google.com/store/apps/details?id=net.routix.mqttdash hl=en`) is a free Android app that makes it possible to create custom user interfaces to control your MQTT devices. To set up a new control surface, you have to enter the destination of your MQTT server, a username, and a password (if you are using a private MQTT server), and then you can start building your user interface. There are various modules that you can add to the surface. For each module, you have various options available:

MQTT Dash: a custom interface with just a few clicks

The only required settings for each user interface element are the name (for example, on/off switch) and the channel that its value should be sent to (for example, `/my-device/power`). Besides these, you can specify the QoS level for the message. Additionally, you can mark it as a retained message. When a client subscribes to a topic, previously retained messages will be received instantly. This has the advantage that the client does not have to wait until a new message is sent to this topic. The following interface elements are available at the time of writing:

- **Switch**: Switches, sometimes called toggles, are great for alternating between two states (for example, on or off). The value that is sent by a switch is either `1` (on) or `0` (off).
- **Slider**: Range sliders allow you to set a minimum and maximum value, and can be used, for example, to set the brightness of a lamp. The value being sent is a number in the range of the minimum and maximum value that you specified.
- **Multiple choice**: Multiple-choice elements are mostly called drop-down elements. You can freely define the items in the drop-down menu; for example, `Mode 1`, `Mode 2`, and `Mode 3`. The value sent when one of the drop-down items is selected is the name of the selected item.
- **Color**: Color pickers are useful for selecting colors; for example, to set the color of an LED strip. The value being sent by this component is the hex value of the color; for example, `#FF0000` (in this case, pure red).
- **Text**: The text input field can be used to enter text; for example, `Hello MQTT`.

The interface of MQTT Dash might seem a bit complicated at first, but once you get the hang of it, it is a powerful tool for prototyping MQTT user interfaces, and it is definitely one of the best apps available for Android.

IoT OnOff (iOS and Android)

IoT OnOff (`https://www.iot-onoff.com`) is an Android and iOS app that is comparable to MQTT Dash. It lets you build a user interface to control your MQTT-powered device(s) and monitor it. It offers many interface elements; for example, switch, time series graph, range slider, color picker, free text, or radial meter. Some of them are purely intended for metering (for example, the time series graph) and for you to build a custom interface to visualize all of the incoming data in a compact format, as well as to control your IoT fleet.

IoT OnOff can either connect to a custom MQTT server, or you can quickly connect to one of the public servers (for example, a public Mosquitto server) that are listed in the app, making it possible to quickly test it out. This image shows the demo interface of IoT OnOff. The individual piece of text is not important and does not need to be read:

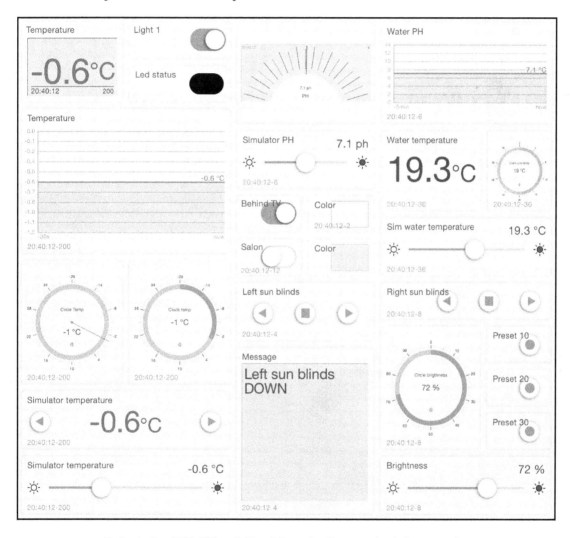

The demo interface of IoT OnOff shows all of the available controls, and is a great starting point for your own project

While IoT OnOff is certainly powerful, the interface is very unintuitive (at least on iOS), and does not follow best practices. But, it is worth the effort, once you find your way around.

Exploring MQTT desktop apps

MQTT clients basically exist for every platform. To publish MQTT messages and subscribe to MQTT topics from your desktop computer, you can choose between many applications. While we will often be using the command-line MQTT client, Mosquitto, throughout this book, feel free to try out a desktop client in parallel. A good MQTT client will offer the same features that Mosquitto has, and can therefore be used interchangeably.

MQTT.fx (Windows, macOS, and Linux)

MQTT.fx (`https://mqttfx.jensd.de/`) is definitely one of the best desktop clients around right now. It is less fancy than the iOS and Android apps that allow you to create your own dashboards via user interface widgets, but it is more solid. The user interface is more organized, and it is a great way to publish and subscribe to MQTT messages. One of the things that I like about it is that it remembers which MQTT servers you connected to before, and which channels you sent messages to. Opening the MQTT.fx app feels like sitting down at your desk, where everything is placed in exactly the same way as you left it the day before.

For `MQTT.fx` to work, you need to have Java installed.

Interface of MQTT.fx

If you struggle to find the text input field in which to place your MQTT payload (as I did), it is the big white rectangle, which takes up most of the app window (where, in this case, **ON** is written).

Understanding MQTT libraries

Due to its open nature, there are a vast amount of MQTT client libraries around. There is a good chance that your favorite programming language offers an MQTT library as well. In the following sections, we will have a short look at MQTT libraries for two programming languages that are relevant to us: Arduino and JavaScript.

Arduino libraries

There are various MQTT libraries around for Arduino, three of the most popular being `arduino-mqtt` (`https://github.com/256dpi/arduino-mqtt`), `pubsubclient` (`https://github.com/knolleary/pubsubclient`), and the Adafruit MQTT library (`https://github.com/adafruit/Adafruit_MQTT_Library`). Some of them are only compatible with certain development boards; also, their implementations of the MQTT specifications differ.

In my projects, I have had the most success with `arduino-mqtt` by Joël Gähwiler, the same person who runs shiftr.io, the MQTT server with built-in graph visualization, which I could not stop praising a few sections prior to this. Because both the MQTT library and the server are developed by the same author, the chances are good that they work well together, and that there are no incompatibilities.

MQTT.js (JavaScript library)

If you want to integrate MQTT into your website, or create a custom app using web technologies that use MQTT to communicate with your Arduino, you need an MQTT library for JavaScript. A library with great support and good development experience is MQTT.js (`https://github.com/mqttjs/MQTT.js`), which makes it easy to publish and subscribe to MQTT messages, and to advance your project with a web frontend.

Summary

In this chapter, we had a look at MQTT—the IoT protocol—which we will use to build the projects in the hands-on part of this book. It is a lightweight protocol, which is especially suitable for the communication of restricted devices, either because of processing power, network bandwidth, or battery power, and therefore, it is a perfect fit for the projects that we are about to build.

MQTT is based on the publish and subscribe pattern. Each message is published to a topic, which can be compared to the categories in a newspaper. A subscriber can subscribe to this information in order to receive updates about it, *"I only want to get updates on foreign affairs and cultural news, but don't care about sports."*

We also learned that a single entity, the MQTT server, manages all of the communication between the clients, and is essential for the communication to work.

Using third-party apps for iOS, Android, or desktop computers, MQTT devices can be controlled, or MQTT messages visualized, without writing a line of code.

In Chapter 4, *Setting up a Lab Environment*, we are going to install all of the tools that we will need for the practical part of this book. Furthermore, we will define which components you need to buy, in order to build the projects.

Questions

1. Who, or what, takes care of all of the MQTT messages that are published by a client?
2. When publishing a message, what are valid topic names?
3. What is the difference between the two wildcards, # and +, when subscribing to multiple topics at once?
4. Can wildcards be used when publishing a message?
5. What is last will (testament)?
6. What are retained messages?

Further reading

- **Official MQTT FAQ**: https://mqtt.org/faq
- **MQTT OASIS announcement**: https://www.oasis-open.org/news/pr/oasis-mqtt-internet-of-things-standard-now-approved-by-iso-iec-jtc1
- **Article about MQTT being used by Facebook Messenger**: https://mqtt.org/2011/08/mqtt-used-by-facebook-messenger
- **Pub-sub client**: https://github.com/knolleary/pubsubclient
- **Emoji table**: https://apps.timwhitlock.info/emoji/tables/unicode
- **List of free MQTT servers**: https://github.com/mqtt/mqtt.github.io/wiki/public_brokers
- **Google Cloud IoT**: https://cloud.google.com/solutions/iot
- **Mosquitto—open source MQTT server to run locally**: https://mosquitto.org/
- **shiftr.io—cloud-based MQTT server and visualization tool**: https://shiftr.io/
- **MQTT Dash—Android app**: https://play.google.com/store/apps/details?id=net.routix.mqttdashhl=en
- **Arduino-MQTT—Arduino library for MQTT**: https://github.com/256dpi/arduino-mqtt

Section 2: Using MQTT in IoT projects

This section includes practical projects, where you will get your hands dirty and apply the concepts you've learned in previous chapters. It includes a preparation section, where we will make sure all of the necessary tools are installed and working, and the main section, where we will build three projects from start to finish. The projects are relatively easy for you to follow and aim to be on point, minimal, but clever, so you're less likely to run into errors while still having something great in your hand at the end.

The following chapters will be covered in this section:

- Chapter 4, *Setting Up a Lab Environment*
- Chapter 5, *Building Your Own Automatic Pet Food Dispenser*
- Chapter 6, *Building a Smart E-Ink To-Do List*
- Chapter 7, *Building a Smart Productivity Cube, Part 1*
- Chapter 8, *Building a Smart Productivity Cube, Part 2*

Setting Up a Lab Environment

4

In this chapter, we will look at the shopping list for all of the components that you will need for the hands-on part of this book, and we will also set up all of the necessary tools and libraries. Furthermore, we will set up an account for the web service, shiftr.io, which can be used for free as a cloud-based MQTT server and visualization tool. We will also explore the instructions for macOS and Windows.

In this chapter, we are going to cover the following topics, which are necessary for working on the hands-on projects in this book:

- The hardware shopping list
- Installing the essential tools and libraries
- Understanding the shiftr.io web service
- Troubleshooting

Hardware shopping list

Before we start coding, you will need to order the necessary parts in order to get going. The components that are listed under **General components** will be needed for all three projects. Please don't forget to order the parts that are needed for the individual projects. They are listed in the following sections of this chapter: *Project 1: a pet food dispenser*, *Project 2: a smart e-ink to-do list*, and *Project 3: a smart productivity cube*.

To reduce costs, I have listed only one Arduino MKR WiFi 1010 development board in the shopping list, which means that you will have to disassemble a project in order to follow along with another one.

You will find an updated shopping list (with links) in the GitHub repository for this book. Instead of searching for the parts manually, have a look at the GitHub repository. Paper books cannot be updated easily after they have been printed, and therefore, for some parts of this book, checking the digital documents on GitHub is recommended (`https://github.com/PacktPublishing/Hands-on-IOT-with-MQTT`). The same is true, by the way, for source code—the most up-to-date version of the code can be found on my GitHub repository.

As you will notice when browsing the parts list, this book is not about building the fanciest Arduino projects. There are many books out there that try to impress; using a multitude of components, and showing off with complex routines. From my experience of teaching electronic prototyping in workshops, it is not about using a lot of different components. The best projects were mostly the simple ones. You need an idea, some key components to make it work, and some artisanal skills to make the final prototype nice to look at. That's it.

The beauty of MQTT is that it is easy to work with, and it is versatile enough to be used in various projects. Powered by Arduino and a few other components, you will be able to bring your ideas to life.

I collected links to SparkFun, an online store for makers, which has a big range of components, tools, and development boards available. Feel free to choose another store, though, as depending on where you live, the shipping costs for packages sent by SparkFun might be expensive and take longer than buying the parts at a local store.

General components

Some components are mandatory for proceeding with the hands-on parts of this book. If you work with electronics, you probably have most of them already. Feel free to reuse whatever you have:

- **Arduino MKR WiFi 1010**: This is the development board that we are going to work with: `https://store.arduino.cc/usa/mkr-wifi-1010`

- **Breadboard**: A half-sized one is needed: `https://www.sparkfun.com/products/12002`
- **Jumper cables (male to male)**: This is the easiest way to connect electronic components to a breadboard: `https://www.sparkfun.com/products/12795`

- **Jumper cables (female to male)**: These are needed to extend the connections of parts that are not directly plugged into the breadboard: `https://www.sparkfun.com/products/12794`
- **Breadboard wires**: These are good for lasting connections on a breadboard. They're not as stable as soldered connections, but are good enough for our prototypes: `https://www.sparkfun.com/products/124`
- **Micro USB cable**: It is important to use a good one, as there are often problems with bad cables: `https://www.sparkfun.com/products/10215`
- **USB power supply (optional)**: You can reuse an existing one, which you probably have anyway for charging your smartphone. You only need an external power supply if you want to use your prototype without a computer powering it; for example, when you want to set it up permanently in your home.

Check out the following video to see the Code in Action:
`http://bit.ly/2oSjufZ`

Project 1 – a smart pet food dispenser

The smart pet food dispenser is a remote-controlled container filled with either pet food, sweets, or cereals. Sending a command from your computer or smartphone will allow you to open and close the dispenser, and make its contents available.

To build this project, you will mostly need household items, which you probably have laying around anyway. The only important electronic component is a servo motor, to open and close the food container.

The following parts are needed:

- **A plastic bottle**: This is a container for the pet food (or cereal if you don't have a pet). Any (big) bottle is fine here. I would recommend a 1.5-liter one, but 1-liter or 2-liters bottle will work equally well. The best are those with a wide opening.
- **Analog servo motor**: This controls the opening of the pet food container. A cheap servo motor is the Tower Pro SG90 Micro Servo. For servo motors, the specified voltage is often 4.8V (instead of 3.3V, as is needed to be used with the Arduino MKR WiFi 1010), but chances are good that it will work just fine using 3.3V. Please make sure that you do not buy a digital servo that is not especially made for 3.3V. Digital ones made for 4.8V will most likely not work with 3.3V: `https://www.adafruit.com/product/169`.
- **Pet food, cereals, or another food to dispense**: In this chapter, peanuts will be used.
- **Tape**: This could be, for example, gaffer tape or duct tape. This will be used to glue together the servo motor and the bottle.
- **Hot glue or superglue**: This is to fix the servo motor arm to the opening cap.
- **A stick or string**: This is used as a fixture for the final construction.

Project 2 – a smart e-ink to-do list

The smart e-ink to-do list contains an e-paper (e-ink) screen, which is very energy efficient compared to regular displays. The device can be hung next to your entry door, to remind you of tasks that you do not want to forget. Every time you walk past the device, you will be reminded of that specific task. You will be able to send commands via your computer, smartphone, or one of the other devices that are built in this book.

The requirements for this project are as follows:

- **E-ink display**: You can use the Waveshare 4.2 inch e-ink display (three-color):
    - `https://www.geekbuying.com/item/Waveshare-4-2-Inch-E-Ink-Display-Module-400x300-Three-color-388289.html`
- **Push button (optional)**: This can be used as a reset button to clear the screen. The nicer-looking buttons that you will typically use when building a case for your prototype require soldering. The following button does not require soldering:
    - `https://www.sparkfun.com/products/10791`

Project 3 – a smart productivity cube

The smart productivity cube is a cube with a custom-built orientation sensor. Using four tilt switches, we will be able to sense whether the cube is resting either on its bottom, top, or any of its other sides. By assigning tasks to each side of the cube, you can use it as a stopwatch for different activities; for example, to time how long you spend learning MQTT, watching TV, or working on project X:

- **Tilt switches**: Tilt switches are very simple components that are handy if you want to sense the orientation of your prototype. Inside the little cube is a metal ball that closes an electrical contact. If you turn it upside down, the ball moves to the other side (due to gravity), and therefore opens the connection. The code that you write when using a tilt switch is exactly the same as for using a normal push button.

  We could also use a more advanced sensor (gyroscope), which can be found, for example, in modern smartphones. Instead of just sensing whether the device is held upside down, they measure the exact angle of orientation. But, in our case, it is enough to know whether the device is held upside down or not. Let's keep it simple. You will need four of these switches, but feel free to order some more. Adding more tilt switches to your project allows you to sense more rotation angles when combined in a clever way. The following links can be used to order tilt switches:

    - `https://www.exp-tech.de/sensoren/sonstige/8555/tilt-sensor-at407`
    - `https://www.sparkfun.com/products/10289`

Optional hardware

The following tools can come in handy when executing the previously mentioned projects, but are in no way required:

- **Multiple breadboards**: The advantage of using one breadboard per project is that you avoid having to completely disassemble a project in order to work on another one.
- **Multiple Arduino MKR WiFi 1010 development boards**: The advantage of using one development board per project is that you avoid having to completely disassemble a project in order to work on another one.
- **Adafruit Perma-Proto boards**: If you want to create permanent prototypes, where you solder the parts together, the easiest way is to use an Adafruit Perma-Proto board (`https://www.adafruit.com/product/571`). It has the same layout as a breadboard, continuous wires on both sides of the board for the power line (3.3V or 5V) and ground.
- **Soldering iron and solder**.
- **Wire for soldering**.
- **Battery for MKR 1010**: For example, this could be a LiPo battery or a battery power pack for smartphones.
- **Timer switch**: This is to save energy. This could be used for the smart to-do list project, in order to save energy by turning it on every x minutes (the e-ink display does not consume energy when not being powered).
- **Toggle switch/slide switch/power switch**: If you equip your projects with an external battery, these will come in handy for opening and closing the power supply, and therefore act as a power switch.
- **Different buttons**: There are many different kinds of buttons available. Some of the tactile push buttons labeled as breadboard-friendly are very hard to keep in place on the breadboard, though. If you don't mind soldering, there are even more to choose from. I have a big collection of buttons that I have assembled over the years, so for every project, I have exactly the right one available. Most of the small push buttons are rather cheap, so it is good to buy a few different sorts.

Installing the essential tools and libraries

In this section, we are going to install all of the tools that we will need in the hands-on part of this book: the Arduino IDE to upload code, Arduino MKR WiFi 1010 board drivers, Visual Studio Code to open non-Arduino code files, and Mosquitto, an MQTT server and toolkit for easy prototyping.

We will also set up an account on shiftr.io, a free MQTT web service, which comes with a useful dashboard.

Additional libraries, which are only needed in one of the hands-on projects, will be set up in that specific chapter.

Installing Arduino

The first application that we are going to install is the Arduino **integrated development environment** (**IDE**). It will be used in all of the hands-on projects in this book to upload code to your Arduino MKR WiFi 1010 (or any other Arduino development board). However, there are other ways to upload Arduino code onto your Arduino, the first being third-party editors, and the second being Arduino Cloud. Third-party editors are preferred by some advanced programmers, who need additional features that the Arduino IDE does not offer (one of them being code completion). The other alternative, Arduino Cloud, also comes with advantages: your code is stored in the cloud; therefore, you can access it from wherever you are. In this book, we are going to use the classic, most simple way, which I personally still prefer over the others—the Arduino IDE:

1. To set up the Arduino IDE, head over to `https://www.arduino.cc/en/Main/Software`, and scroll down a bit, until you see the download section.

2. Download the Arduino IDE, and click the **Download** button for your platform. Use the **Windows Installer** if you are using Windows or **Mac OS X** if you are on macOS. If you are on Windows, please make sure that you download the Windows Installer, and not the ZIP or app version. These have different file paths. During the rest of the installation instructions, I refer to these file paths, so you would have to figure things out by yourself if you chose the ZIP or app version. For macOS and Windows, please also make sure that you download the latest stable version (not an hourly or beta build, which can be less reliable):

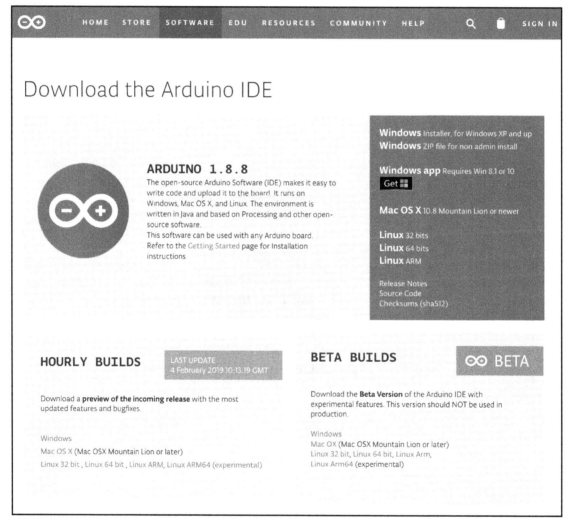

The Arduino IDE download page

At the time of writing this book, Version 1.8.8 was the most recent one. But when you read this, a more recent version will probably be available.

3. Before the download starts, you will be asked whether you want to contribute to the Arduino Software by making a donation:

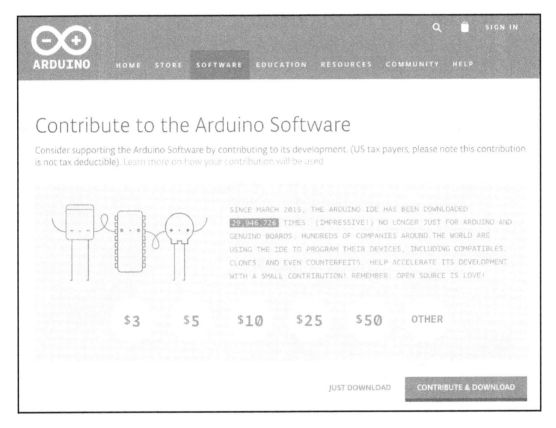

Arduino contribution page

While it is definitely appreciated by the Arduino team, you don't have to. You can click **JUST DOWNLOAD**.

In the next section, we will install Arduino on your platform.

Installing Arduino on macOS

In the previous section, you downloaded the right version of Arduino for macOS. After the download has finished, extract the ZIP file, and drag the extracted application, `Arduino.app`, from the download folder to your applications folder.

Installing Arduino on Windows

After the download of the Arduino Windows Installer has finished, follow the succeeding steps to install Arduino on Windows:

1. Run the `.exe` file.
2. You will be asked by Windows whether you trust this application, and whether it is allowed to make changes to your computer. Here, you can safely press **Yes**.
3. Agree to the license agreement: click **I Agree** (you might want to read it first, but that's your decision).
4. In the next step, additional features can be selected. Just leave them as they are (in my case, all boxes ticked) and click **Next**:

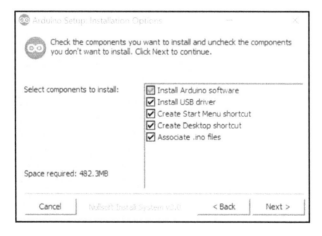

The Arduino IDE installation options

5. You will now be asked in which folder Arduino should be installed. Just use the default choice here and press **Install**.
6. You will be asked (maybe even multiple times) whether you want to install new drivers. Just press **Install**, and leave the **Trust Adafruit** checkbox checked.

7. When the setup is done, you can close the setup window.
8. Launch the Arduino IDE. When you launch it for the first time, you will be asked whether you grant the Arduino Software access to your network(s). You can activate both check marks, and click **Grant access**.

Next, we will install Arduino MKR WiFi 1010.

Installing Arduino MKR WiFi 1010

I hope your order of the components and Arduino development board has arrived by now. But don't connect your new Arduino MKR WiFi 1010 just yet. We first need to set up the board in the Arduino IDE. Follow these instructions:

1. Launch the Arduino app, and navigate to the **Tools** menu, then click on **Boards**, and then **Boards Manager**.
2. Search for the name of the board that you want to install; in our case, Arduino MKR WiFi 1010. The first result that will show up is **Arduino SAMD Boards**—this unifies the Arduino drivers of various boards, which are based on the SAMD microprocessor.
3. Click on **Install** to start the installation of the Arduino SAMD Boards package. The version that is selected by default is fine, and it will install the latest version:

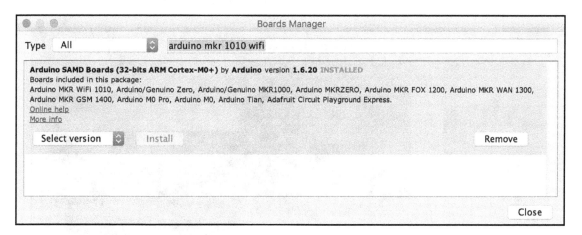

The Arduino Boards Manager

Because I have already installed the MKR WiFi 1010 board, in the preceding screenshot, you see a **Remove** button. Yours will be an **Install** button, instead.

Depending on which operating system you are using, there might be additional dialog boxes:

- **A note for macOS users**: The official documentation (see `https://www.arduino.cc/en/Guide/MKRWiFi1010`) also mentions that you might get asked whether you want to open **Network Preferences**, and that you should just click **Apply**. Even when the Arduino MKR WiFi 1010 shows as not configured in the **Network Preferences**, it is ready to use.
- **A note for Windows users**: You might be asked whether you trust the developer during the installation of the board. They can be trusted. Click the **Allow** or **Trust** button, and continue.

After you plug in the Arduino MKR WiFi 1010, Windows might initiate the installation of system drivers, which are needed in addition to the board we installed in the Arduino IDE.

Testing your Arduino

You can now connect your **Arduino MKR WiFi 1010** via USB with your computer. Before we can run our own programs (also referred to as sketches) on the Arduino, we need to make sure that the correct board is selected:

1. Go to **Tools** | **Board** and select **Arduino MKR WiFi 1010**:

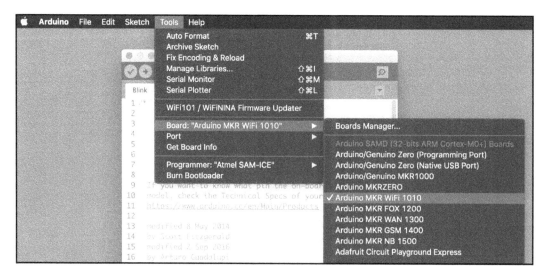

The Arduino IDE board selection menu

2. We also need to make sure that the correct port is selected, so that the Arduino IDE knows where to upload your code to. Go to **Tools** | **Port**:

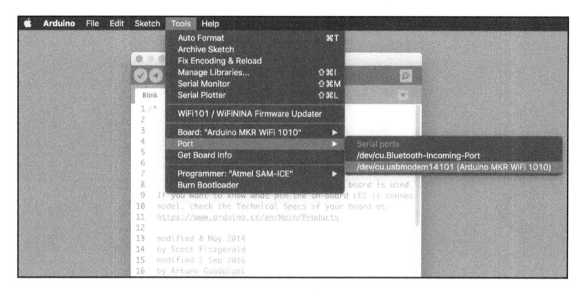

The Arduino port selection menu

3. On Windows, you will see an entry called COM3 (Arduino MKR WiFi 1010) (or similar); select it. On macOS, the correct port to choose will be named /dev/cu.usbmodem14101 (Arduino MKR WiFi 1010), or similar.

> If it is less obvious which port is the correct one (you don't see **Arduino MKR WiFi 1010** mentioned in the list), a good trick is to disconnect the USB cable from your computer, remember which entries are available in the **Ports** list and then plug the USB cable with the Arduino MKR WiFi 1010 back in. The new port that shows up is the correct one.

4. Now that we have selected the correct board and port, we are all set up to try out our first example. Go to **File** | **Examples** | **01.Basics** | **Blink**. This example switches the built-in LED of the Arduino on and off. This is probably exactly what your Arduino has been doing from the beginning, after you plugged it in, since it comes pre-flashed, with this example firmware.

Let's make a small adjustment to the code, to see whether everything is working. Follow these steps to implement the changes:

1. Scroll down until you see the following code:

```
void loop() {
    digitalWrite(LED_BUILTIN, HIGH); // turn the LED on (HIGH
is the voltage level)
    delay(1000); // wait for a second
    digitalWrite(LED_BUILTIN, LOW); // turn the LED off by
making the voltage LOW
    delay(1000); // wait for a second
}
```

The time unit of the delay function uses milliseconds, so 1,000 equals 1 second.

2. Change the first (or second) `delay(1000)` to `delay(100)`, and press the upload icon (the second button in the menu bar, with the arrow button pointing to the right):

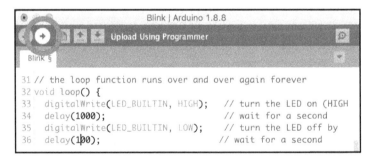

The upload button

3. While uploading the sketch, the onboard LED of the Arduino will go on and off rapidly, indicating that new firmware is being uploaded.
4. After a few seconds, the code will be uploaded, and you now should see a different LED flashing pattern than before (the LED should be on for a shorter time now, with the off time staying the same).

Even after many years of programming microcontrollers, I keep coming back to the blink sketch to verify that my development board is working as intended. When encountering problems, you need to figure out what definitely works, and what possibly does not, in order to isolate the cause of an error. Flashing my development board with the blink sketch lets me know that code can be uploaded successfully to the board, and that it is not completely broken (many people refer to this state as bricked, as in: as useful as a brick).

Installing the WiFiNINA library for Arduino

In order to connect to the internet, your Arduino MKR WiFi 1010 needs the WiFiNINA library. To install the library, first open the Arduino IDE, then follow these steps:

1. Open the library manager by clicking on **Tools** | **Manage Libraries**.
2. Search for `WiFiNINA`, and install the most recent version of the library:

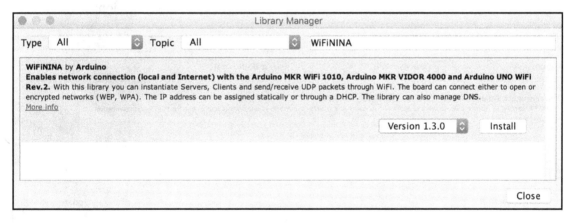

The WiFiNINA library in the Arduino Library Manager

Now, there is only one library missing that we need in all of the hands-on chapters: a library to use MQTT on the Arduino MKR WiFi 1010, which we will install in the next section.

Installing the MQTT library for Arduino

To use MQTT on the Arduino MKR WiFi 1010, we need to install one of the available MQTT libraries. We will use `arduino-mqtt` (`https://github.com/256dpi/arduino-mqtt`) for this, an MQTT library made by Joël Gähwiler, who also made shiftr.io (`https://shiftr.io`), the MQTT server and visualization platform that was introduced in `Chapter 3`, *Getting Started with MQTT*.

Follow these steps to install the MQTT library:

1. Click on **Tools** | **Manage Libraries**.
2. Enter `arduino mqtt` in the search bar.
3. Press *Enter*, and then scroll down until you find a library called MQTT, by Joël Gähwiler.
4. Install it by clicking on the **Install** button on the right. I am using Version 2.4.3. You can install the latest version here, as long as it is a 2.x.x version:

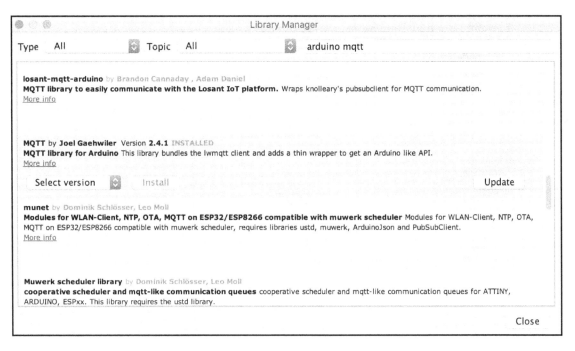

The Arduino Library Manager

Before we continue with the installation of the required tools, we will have a quick look at what the version numbers actually mean in the following section.

Understanding semantic versioning

Many library authors use a versioning scheme called **semantic versioning** (`https://semver.org/`), which makes it easier for developers like you and me to update their libraries, without having to change their code, and without the fear that your program will not work any more after an update. Software (in this case, the MQTT library) using semantic versioning always follows the same schema:

```
Major.Minor.Patch (for example 1.2.3)
```

Here, `1` is the major version, `2` the minor version, and `3` the patch version.

New versions coming out that fix tiny issues (called patch releases) lead to an increased patch version number, for example, `1.2.3 | 1.2.4`.

When new features are added, which are non-breaking (the old code will still work), the minor version is increased, for example, `1.2.3 | 1.3.3`.

When there are some big changes, which lead to breaking changes (so old code that uses the library might not work any more), the major version is increased, for example, `1.2.3 | 2.0.0`.

But, as with most things in life, there is an exception: in the initial development phase of software/libraries that are versioned using semantic versioning, the major version, 0, is used, so for example, 0.1.0. When the major version is 0, breaking changes might appear even with minor or patch version increments. So, if you start using a library with the version number 0.1.0 and update it, because a new version 0.2.0 was released (the Arduino IDE will let you know about it), your existing code that uses the library might not work any more. Mostly, it will anyway, but sometimes, updating the library might be the problem, because the way the library should be used has changed, and therefore your existing code might break.

Long story short, when library authors follow this way of versioning, which the MQTT library that we are using does, you can safely install any 2.x.x version, and it will work in the same way as in the code in this book, which uses Version 2.4.3. There might be some new features and bug fixes, but the code that uses the library will look and work in the same way.

Before we get back to our topic, let's talk about MQTT version numbers a bit more. The MQTT protocol also follows semantic versioning. As we discussed in Chapter 3, *Getting Started with MQTT*, there are two MQTT versions that you should know about: MQTT 3.1.1, which is widely supported as of early 2019, and Version 5, which is being adapted more and more, but the support in third-party libraries and MQTT servers is far behind the support for the 3.1.1 specification.

MQTT 3.1.1 has a major version of 3. The new version getting traction at the time of writing, 5, has a different major version number. Therefore, code written for MQTT 3.1.1 is not completely compatible with, for example, a server that is running MQTT Version 5.x.x (in case this is not clear, x stands for any number).

Installing Visual Studio Code

Next, we need a code editor, which is useful when working with Markdown files or HTML, CSS, and JavaScript code. For example, if you want to create a web front-end to go along with your physical prototype, you can use a code editor to build it. There are many different editors to choose from. Some of the most popular ones are Visual Studio Code, Atom, and Sublime Text. But, if you are familiar with another one, feel free to keep using it.

All of the three mentioned code editors are excellent choices for web development. Visual Studio Code has the best community support though, it responds very quickly, and it is, in general, a breeze to work with.

Now, go to the download page of Visual Studio Code: `https://code.visualstudio.com/download`:

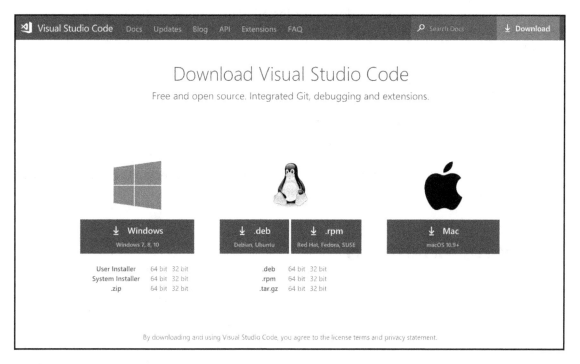

The Visual Studio Code website

Installing Visual Studio Code on macOS

Download Visual Studio Code for Mac. After downloading has finished, extract the ZIP file, and drag the extracted application, `Visual Studio Code.app`, to your `applications` folder.

Installing Visual Studio Code on Windows

On Windows, you should download the user installer; in most cases, the 64-bit version. If you are not sure whether you're running a 32- or 64-bit operating system, you can have a look in the system settings:

1. Click on the Windows **Start** button.
2. Navigate to **Settings** | **System** | **About**.
3. In **Device specification**, you can see whether you are running a 32- or 64-bit operating system.

Download the user installer that fits your operating system and execute the `.exe` file when done. Follow the installation instructions from the installer.

Installing Mosquitto

Now, let's install Mosquitto, a very popular open source MQTT server that you can run locally on your computer. It is great for simple testing, and I use it regularly when developing MQTT applications, before eventually switching to a cloud solution.

Installing Mosquitto on macOS

To set up Mosquitto, we first need Homebrew (or brew for short), a very popular package manager that is often the preferred way of installing command-line tools on macOS. Unlike the myriad of development tools related to coffee, this one is related to beer. Cheers!

Now, follow these steps to start the installation:

1. Go to `https://brew.sh`, copy the install command shown on the website, and paste it into Terminal. The command will look something like this:

   ```
   /usr/bin/ruby -e "$(curl -fsSL
   https://raw.githubusercontent.com/Homebrew/install/master/insta
   ll)"
   ```

As there might have been changes in the installation method, you should copy and paste the command that is listed on the website instead of re-typing the preceding one.

2. After Homebrew is set up correctly, we can install Mosquitto by running the following command in the Terminal:

    brew install mosquitto

3. This might take a while, as Homebrew will make sure that it is up to date before installing any packages. When it is done, you will be presented with some text output. One of the important bits is this:

    ```
    Mosquitto has been installed with a default configuration file. You
    can make changes to the configuration by editing:

    /usr/local/etc/mosquitto/mosquitto.conf

    If you need to fine-tune Mosquitto, this is the file to do the
    changes in. But you probably won't need to do any changes, and you
    definitely should not change anything without knowing what it
    really means.
    ```

4. There is one thing left to do in order to make Mosquitto work. In the Terminal output, we can also see the following:

    ```
    To have launchd, start Mosquitto now and restart at login:
    ```

    brew services start mosquitto

5. As a final step, we therefore need to make sure that Mosquitto is started automatically, by running the following command:

    brew services start mosquitto

That's it. We are all set!

Installing Mosquitto on Windows

Follow these steps to set up Mosquitto on Windows:

1. Download the Mosquitto installer from the official Mosquitto website (`https://mosquitto.org/download/`). As we discussed before, in the section on how to install Arduino, you need to pick the right version for your version of Windows—either 32 bit or 64 bit. In most cases, you will need the 64-bit version. In my case, the latest version available is `Mosquitto-1.5.6-install-windows-x64.exe`.

2. Once downloaded, run the installer and follow the instructions. Windows might try to protect you and block the execution, claiming that the installation would pose a risk to your computer. Mosquitto is developed by a trusted entity (the Eclipse Foundation), and therefore it is not a risk. Click on **More information**, which will reveal another button: **Execute anyway**. Click it to proceed with the installation. When asked by Windows whether you want to allow this application by an unknown developer to be installed, click **Allow**.

3. Once the installer starts, you need to press **Next** a couple of times, using the choices that are pre-selected. Once you reach a screen where you see the **Files** and **Services** checkboxes, make sure both are checked (default). Mosquitto will set up a background service, which runs a Mosquitto server on your computer. Similar to a web-based MQTT server, you can connect to it, and send and receive messages.

There is one thing left to do to—make it more convenient to use the Mosquitto commands. By default, the Mosquitto commands can only be used in the directory in which Mosquitto was installed, which means that every time you want to use these commands, you first have to navigate to this directory in PowerShell. Tedious. Let's fix that:

1. Open PowerShell (by clicking the Windows key, then entering `PowerShell` and pressing *Enter*). Then, navigate to the directory in which Mosquitto was installed by issuing the following command (followed by *Enter* to execute it): `cd "C:\Program Files\mosquitto"`. On a side note, `cd` is short for change directory.

2. Now, we could run the two commands we are about to use: `mosquitto_sub` to subscribe to messages, and `mosquitto_pub` to send messages. But to do so, we would have to navigate to the installation directory of Mosquitto, each time using the `cd "C:\Program Files\mosquitto"` command. There is an easier way. Let's create two aliases by running the following command:

```
Set-Alias mosquitto_pub "C:\Program
Files\mosquitto\mosquitto_pub"
Set-Alias mosquitto_sub "C:\Program
Files\mosquitto\mosquitto_sub"
```

That's it! Now, you can just start PowerShell and use `mosquitto_sub` and `mosquitto_pub`, without needing to use `cd`.

Testing Mosquitto

Let's see whether everything works as expected. Open PowerShell on Windows/Terminal on macOS. Follow these steps to test Mosquitto:

1. Enter `mosquitto_sub -t "/test"` to subscribe to the `/test` topic (and therefore get notified whenever there is a new message in the channel).
2. Now, open another PowerShell/Terminal window, and send a message to this channel by running this code:

```
mosquitto_pub -t "/test" -m "Hello!"
```

The `-t` stands for *topic*, and the `-m` for *message*. So, if you were to put it in a sentence, it would mean: Publish the «Hello» message to the «/test» topic.

In the other PowerShell/Terminal window, you should now see the message: `"Hello!"`. Hooray, our Mosquitto installation has worked!

Understanding the shiftr.io web service

shiftr.io is a unique web service that makes working with MQTT a breeze. Using a graph-based real-time visualization, debugging MQTT messages becomes very easy:

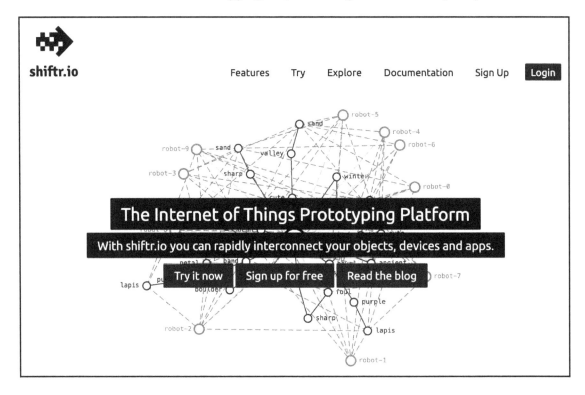

The website of shiftr.io—the network in the background is an actual representation of an MQTT network; the text and numbers in this image are intentionally illegible

As you have probably noticed in previous chapters, I am not a huge fan of proprietary, closed web services, mainly because of the lock-in effect. If you create an application that uses a certain web service, in most cases, you are tied to it. Your code will only work if the web service is online, and if you decide to move to an alternative web service provider, you often have to put in a lot of effort to make your code work on the other platform. For platforms that are built around MQTT, this effort is minimized. If, one day, the provider decides to cut loose the free account that you might be using, increase costs, or simply close down, you can move on to the next alternative that provides an MQTT server. If the provider implements the MQTT specification correctly, and is based on the same version of MQTT (currently, the most-used version is 3.1.1), the effort will be minimal. In a best-case scenario, you just replace the MQTT server URL and your login credentials, and that's it. A long list of free MQTT servers can be found on the official MQTT wiki in the GitHub repository (`https://github.com/mqtt/mqtt.github.io/wiki/public_brokers`).

But shiftr.io does not seem to be going anywhere. It's quite the opposite. As you can see on the official blog (`https://blog.shiftr.io/`), there are recent updates to make it even more useful, for example, using private namespaces, or time series charts.

But enough of the praise. Before being able to use it, you need to create an account. Go to the sign-up page and create a (free) account: `https://shiftr.io/sign-up`.

All of the features that we are about to use are currently free.

After signing up, you might need to click a confirmation link that is sent to your email. Log in to your newly created shiftr.io account and you will see your personal dashboard, which, at the beginning, is empty.

In the following screenshot, you can see what my dashboard looks like. I have used shiftr.io for various projects, and I can instantly see which of these projects are currently activated (sending messages), what my topic structure looks like, and which devices are subscribed to my channels:

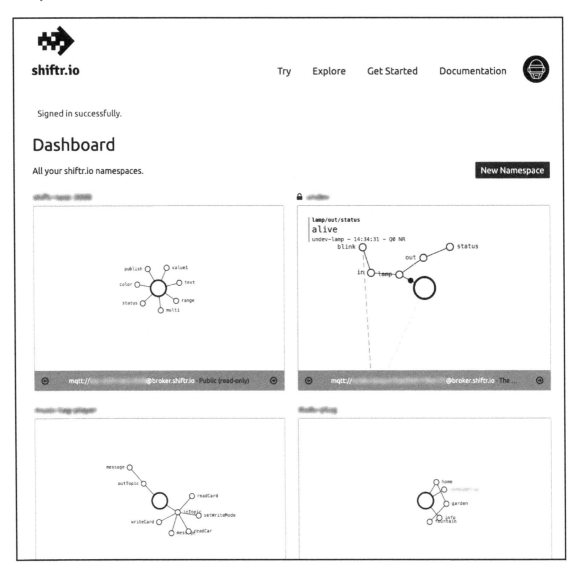

This is an image of the personal shiftr.io dashboard; the text and numbers in this image are intentionally illegible

For every project on the shiftr.io dashboard, you see a widget showing you which messages are sent, what your topic structure looks like, and which devices are connected to this specific namespace.

While being based on MQTT, shiftr.io brings some unique features that are not part of the MQTT standard.

Namespaces in shiftr.io can be seen as mini-MQTT servers. In the preceding screenshot, you can see that each of the four projects that are currently visible form a mini-network—consisting of clients that connect to it, and topics where messages get sent and which clients it can subscribe to. This makes it pseudo-private. You have an isolated graph, which only shows what is relevant for it. Under the hood, these are public, though, and can be accessed by other people as well. Other users see all of the messages that are sent to a public namespace:

```
/your-namespace/living-room/funky-arduino-project
/your-namespace/living-room/smart-lamp
/another-namespace/kitchen/coffee-machine
/another-namespace/basement/alarm
```

If you look at your graph, you will only see your namespace, with the namespace part stripped away:

```
/living-room/funky-arduino-project
/living-room/smart-lamp
```

While the representation looks like the messages are private, by default, they are not. But, for most prototyping needs, this is fine. Since mid-2018, shiftr.io also offers private namespaces, which offer more privacy and access control. When you decide to work on a critical project, let's say an MQTT-based smart door lock (don't do that please), using a public namespace would not be a good idea.

But now, let's actually use shiftr.io. There are basically two ways to use shiftr.io. Without an account, you can still use the public test namespace (`https://shiftr.io/try`). Because many people try out shiftr.io, it is quite messy there, because there are so many messages sent around, so it is hard to see which messages were sent by us.

Let's create our own namespace:

1. Click on your user avatar in the top-right corner and select **New namespace** (you can also reach it via `https://shiftr.io/new`).

2. On the following page, enter a name for your namespace (this needs to be unique for your user account), and a description. For now, we will be using public namespaces, so you can leave the **Private** checkbox disabled:

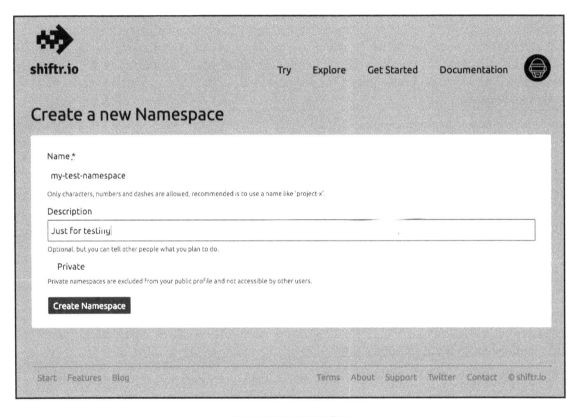

The new namespace page on shiftr.io

3. After you create your first namespace, you need to create a token, so you can send messages to your namespace (`write-access`). Click on **Namespace Settings**:

shiftr.io namespace page

Later on, after sending a few messages to your namespace, you will see a graph on this page visualizing your connected clients and all of the topics that have received messages.

4. In your namespace settings, you will now see an (empty) list of tokens. We haven't created one yet, so let's do that. Click on **Add token**:

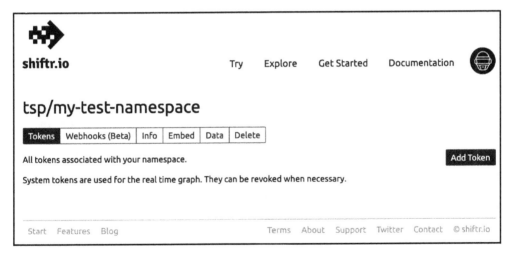

shiftr.io namespace settings

5. In the new token dialog, you can define a name for your token (this needs to be unique on the whole shiftr.io platform), a password, and a description, so later on, you will have an idea why you created it and what it is being used for:

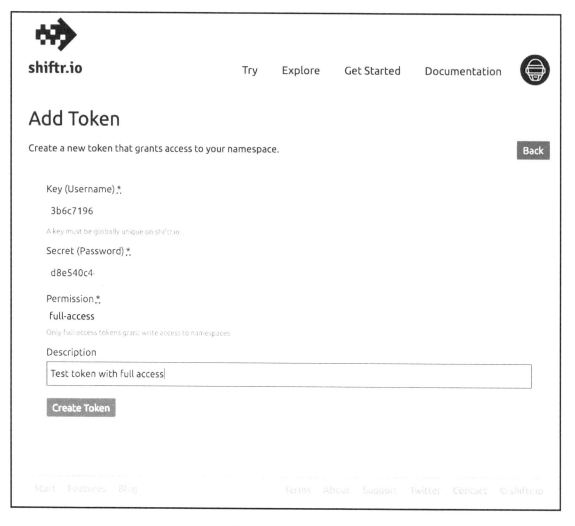

The new token dialog on shiftr.io

Great! We are now ready to use our token to send messages to the MQTT server. Please note that parts of my password are blurred, so my private password remains private, but you can still see where it is being used:

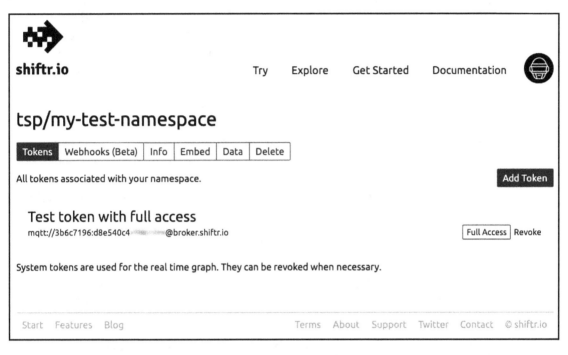

shiftr.io namespace settings with a full-access token

6. Now, open Terminal (macOS)/PowerShell (Windows) and paste the following command:

```
mosquitto_sub -t "/test" -h broker.shiftr.io -u 3b6c7196 -P
d8e540c4********
```

This means: subscribe to the `"/test"` topic on the MQTT server, `broker.shiftr.io`, with the `3b6c7196` username, and the `d8e540c4********` password. Replace the username and password with your private ones from the token that you have just created.

In case you are wondering why you have to enter all of this into Terminal/PowerShell, and when will we finally start using MQTT together with Arduino: using the Terminal is a great way to try out the features of MQTT, define a topic structure, and so on. If your project grows in complexity, you will save a lot of time when you prototype MQTT outside of the Arduino world, before you actually code and upload it onto the Arduino.

If you run the preceding command, nothing will happen. Well, at least, you should not see anything. In this case, seeing nothing is good—no error message:

1. Go back to the details page of your namespace by clicking on the namespace name at the top; in my case, this is `my-test-namespace`. You should now see another circle in the graph—yay! This is the connection coming from your Terminal. But there is just a connection and no messages yet.

2. Open another Terminal/PowerShell window and, this time, send a message instead of subscribing. Do you have an idea how to do it? You need to combine the `mosquitto_pub` command that we used earlier, and add the parameters of our most recent `mosquitto_sub` command, so that we use `broker.shiftr.io` as our MQTT server, together with our username and password. Also, we need to add a message to it, by adding `-m "Hello"` to the command. It does not matter here, in what order the parameter pairs are used.

3. Reorder them however it makes sense to you. For me, having the topic and the message at the beginning gives me the best overview:

```
mosquitto_pub -t "/test" -m "Hello" -h broker.shiftr.io -u
3b6c7196 -P d8e540c4******
```

Try to write it yourself instead of just copying and pasting the command. You won't save too much effort anyway, because you need to replace the username and password with your personal credentials.

If you now go back to your namespace dashboard, you will see our message in the top-left corner (it always shows the last message that was sent, here).

If you have the browser tab and the Terminal tab open at the same time, you will notice that you even see a neat animation in the shiftr.io graph whenever a message is sent. Try sending a few more messages.

One thing we can improve on is specifying a client ID. If you open even more Terminal tabs, or, later on, have a lot of Arduinos communicating, you need to identify which one sent which message. In Mosquitto, we can use the `-i` flag for this in order to provide a client ID (`-i` for client ID is not super easy to remember, but think of `i`, as in, ID). Here, I am using `"second-tab"` as the name of the publisher:

```
mosquitto_pub -t "/test" -m "Hello" -h broker.shiftr.io -u 3b6c7196 -P
d8e540c4****** -i "second-tab"
```

On the shiftr.io namespace page, you can now see our message, `test`, in the top-left corner:

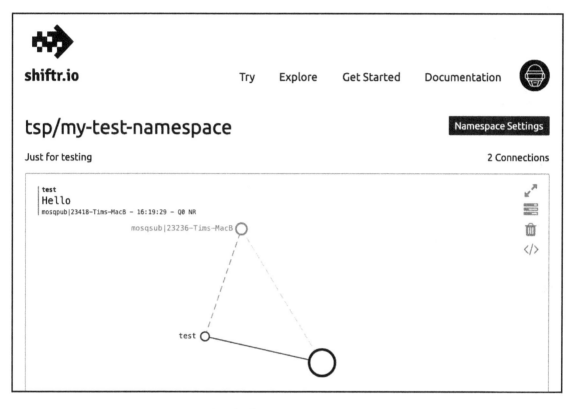

Visualisation of a private MQTT network on shiftr.io

I hope you did not have any problems following along so far and are excited to finally make use of MQTT together with Arduino in the next chapter.

Troubleshooting

If you had any problems with setting up one of these tools, or if you run into any problems during the projects in this book, feel free to open an issue in the repository for this book on GitHub (`https://github.com/PacktPublishing/Hands-on-IOT-with-MQTT/issues`), and I will do my best to help you to fix it. Installing Mosquitto especially can be tricky at times, even more so on Windows.

Don't be shy! If you have a problem, chances are that another reader also has the same problem. You could find it in the issue list and be presented with a solution. If you do not find a solution to your question, just create a new issue. You will need a GitHub (`https://github.com`) account to do so.

Also, please have a look at the main page of the repository (`https://github.com/PacktPublishing/Hands-on-IOT-with-MQTT`). Here, you will find all of the updated information in the repository, as books are hard to correct once printed.

Summary

The purpose of this chapter was to prepare you for the following hands-on chapters, by giving you a list of materials that you will need, as well as installing all of the tools that we are going to be using.

Together, we set up an account on shiftr.io, the MQTT web service that we will be using throughout this book. We created a private namespace, together with a token. This makes it possible for you to have your own MQTT area, to which only you can publish MQTT messages. On the shiftr.io website, you can inspect all of the devices that are connected to your MQTT network, and the messages that are being sent to your namespace. You also learned how to deal with the problems that you encounter on the way, and where to look for solutions.

The knowledge that you have acquired in this chapter will help you immensely in the hands-on projects that we will explore in the subsequent chapters. In Chapter 5, *Building Your Own Automatic Pet Food Dispenser*, we will get our hands dirty and finally create our first MQTT-powered project—a smart pet food dispenser.

Questions

1. How do you send a message using Mosquitto?
2. How do you subscribe to messages using Mosquitto?
3. What does –t stand for when using the Mosquitto commands?
4. What does –m stand for when sending messages via MQTT?
5. Have a look at the explore page of shiftr.io (`https://shiftr.io/explore`). Here, you will find public namespaces from other people. How do they use shiftr.io and MQTT? Can you tell by looking at the namespaces, topics, and connected clients? How do other people organize their topics? Which structure do they use?

6. Try opening a few other tabs and sending messages from each of them. Open another one and experiment with subscriptions. Do you remember how they work? You can use slashes to create sub-topics (for example, /test/sub-topic), or use wildcards to subscribe to multiple topics at once (for example, /test/#), which will get an update once something in /test or a sub-category of /test has been published.

7. Try and see whether you can figure out how to use an MQTT app for Android or iOS to work with your namespace on shiftr.io. Basically, you have to enter a username, a password, and a server URL, in the same way that we just did using mosquitto_pub and mosquitto_sub.

Further reading

- **Official getting started guide for the Arduino MKR WiFi 1010**: https://www.arduino.cc/en/Guide/MKRWiFi1010
- **Arduino download page**: https://www.arduino.cc/en/Main/Software
- **Homebrew, a package manager for macOS**: https://brew.sh/
- **Mosquitto download page**: https://mosquitto.org/download/
- **Visual Studio Code download page**: https://code.visualstudio.com/download
- **List of public MQTT servers**: https://github.com/mqtt/mqtt.github.io/wiki/public_brokers
- **shiftr.io getting started guide**: https://shiftr.io/get-started

5
Building Your Own Automatic Pet Food Dispenser

It is finally time to get your hands dirty! In this chapter, we will build an automatic pet food dispenser. It will hold one portion of pet food and can be controlled from anywhere in the world via MQTT, including one of the many available MQTT apps for Android or iOS.

The smart pet food dispenser is easy to build and will show you how to create smart devices with only a little effort. Just by using a servo motor, there are a lot of different smart devices that you will be able to make after following along with this chapter.

If you don't have a pet, you might want to build yourself the first part of a breakfast robot—a cereal dispenser. The setup and code will be exactly the same. You can decide what the dispenser should hold—maybe sweets or bubblegum!

The following topics will be covered in this chapter:

- Testing the components
- Controlling the servo motor via the Serial Monitor
- Building the smart pet food dispenser
- Optimizing the dispenser code
- Improving the visual appearance
- Making the dispenser controllable via MQTT

Technical requirements

Please make sure you have completed `Chapter 4`, *Setting Up a Lab Environment*, and installed all of the necessary tools and libraries before continuing with this chapter.

The installations required for this chapter are listed here:

- **Arduino IDE**: Editor to program the Arduino
- **Arduino MKR WiFi 1010 board driver**
- **Mosquitto**: Command-line MQTT client and server
- **WiFiNINA**: Arduino library to be able to use the Wi-Fi chip on your Arduino MKR WiFi 1010
- **Arduino MQTT**: Arduino library for MQTT

The following electronic components are needed:

- **Arduino MKR WiFi 1010**: This is the development board we will be using: `https://store.arduino.cc/usa/mkr-wifi-1010`
- **Micro USB cable**: This is to connect the Arduino to your computer: `https://www.sparkfun.com/products/10215`
- **Jumper wires**: This is to connect the servo cable to the Arduino: `https://www.sparkfun.com/products/12795`
- **Analog servo motor**: This controls the opening of the pet food container. A cheap servo motor is the Tower Pro SG90 Micro Servo. For servo motors, the specified voltage often is 4.8V (instead of 3.3V, as needed to be used with the Arduino MKR WiFi 1010), but chances are that it will work just fine using 3.3V. Please make sure you don't buy a digital servo that is not especially made for 3.3V. Digital ones made for 4.8V will most likely not work with 3.3V:
  - `https://www.adafruit.com/product/169`

You can find the code for this chapter in this book's GitHub repository (`https://github.com/PacktPublishing/Hands-On-Internet-of-Things-with-MQTT`) in the `ch5` folder. Please try to write the code yourself and only use the code from the repository if you run into problems. This way, you will learn the most.

Check out the following video to see the Code in Action:
`http://bit.ly/2oSjufZ`

This project is minimal in terms of the components being used. This is by design. I don't think you will learn much by creating a super-complex project by copying and pasting a lot of code into your Arduino IDE. Instead, I want to show you as little code as possible, making sure you understand it, and inspire you to think about other project ideas that can be brought to life with just a few components. There are so many possibilities that do not require a big budget or lots of electronic components, especially if you think about the possibilities with an internet-connected servo motor. I am sure you can come up with many more project ideas.

If you run into any problems, feel free to create an issue on GitHub in this book's repository. Chances are that you'll either find a solution to your problem there or if you do not find an answer to your question, another reader might profit from your question.

Some additional requirements

You will need the following household items for this project:

- **A plastic bottle**: This will be the container for the pet food (or cereal if you don't have a pet). Any (big) bottle is fine here. I would recommend a 1.5 l one, but 1 l or 2 l will work equally well. The best are ones with a big opening.
- **Pet food, cereal, or other food to dispense**: In this chapter, peanuts will be used.
- **Tape**: You could use gaffer tape or duct tape. This will be used to glue together the servo motor and the bottle.
- **Paper or cardboard**: For the opening cap as well as to make the prototype look nicer, we will use thick paper or thin cardboard.
- **Hot glue or super glue**: This will be used to fix the servo motor arm to the opening cap.

Testing the components

A good workflow when working with microcontrollers is to verify that the components in your project work individually before you combine them. One of the mistakes made by many Arduino beginners is connecting too many things at once. This makes it very hard to diagnose the cause of any problems and, in turn, makes it difficult to solve them.

A better approach is to spend a little bit more time up front and make sure every component you are planning to use in your project works on its own. When you, later on, attach multiple components and write more complex code, it will be easier to debug.

So, instead of jumping right into our main code, we will start small and verify that the different components of our smart pet food dispenser work. We will verify three things:

- Check whether the Wi-Fi is working
- Check whether MQTT is working
- Check whether the servo motor is working

 When working on more complex projects with multiple electronic components, it can be helpful to use multiple breadboards to test components individually. You can, for example, have one test circuit with just an LED and one with just a servo motor. By switching your development board between the two breadboards, you can make sure the stepper motor works individually, as well as the LED. Then, you can combine the two sketches into one using both components, while being sure that the wiring is correct and the code for the individual components works as well.

Let's start testing our Wi-Fi connection on the Arduino MKR WiFi 1010. Once we have completed all three mini-tests, we will build upon it and create our main code.

Checking Wi-Fi connectivity with Arduino MKR WiFi 1010

When working on a project that depends on internet access, one of the first things to verify is that your Arduino MKR WiFi 1010 can connect to the internet. In `Chapter 4`, *Setting Up a Lab Environment*, we installed the WiFiNINA Arduino library. We need this library to make use of the Wi-Fi capabilities of the Arduino MKR WiFi 1010.

To verify that Wi-Fi is working on the Arduino MKR WiFi 1010, we can use the WiFi Ping example, which is part of the WiFiNINA library. Follow these steps to verify that Wi-Fi is working:

1. Open the **WiFiNINA** | **WiFiPing** example. This example is a good way to verify that the Arduino can connect to your local wireless network and that it can access other servers on the internet:

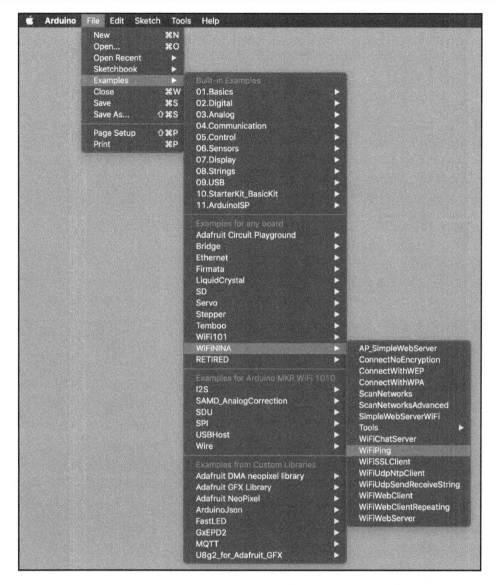

WiFiNINA Ping example

2. After you have opened the example, navigate to the second tab
   (arduino_secrets.h), where you will find a define statement for the network
   username and password, which we need to change:

```
#define SECRET_SSID ""
#define SECRET_PASS ""
```

3. Now, enter your username and password in the quotation marks, which you will probably find on a sticker on your network router. After entering the username and password, your code should look something like this:

```
#define SECRET_SSID "Your Wifi Name"
#define SECRET_PASS "1234-TOP-secret-P455W0RD"
```

4. Press the **Verify** button in the top-left corner of the Arduino IDE (the button with the checkmark) to make sure there are no syntax errors. As we didn't change much, this should result in `Done compiling` being printed in the bottom part of Arduino IDE without showing any errors.

5. Now, let's upload the code by pressing the **Upload** button (the arrow pointing to the right), which is next to the **Verify** button.

The output should look something like this:

```
[=====================  ] 73% (192/263 pages)
[============================ ] 97% (256/263 pages)
[============================] 100% (263/263 pages)
done in 0.109 seconds
Verify 16808 bytes of flash with checksum.
Verify successful
done in 0.015 seconds
CPU reset.
```

It looks like it worked, but to find out for sure, we need to check the serial output of the Arduino:

1. Click on **Tools | Serial Monitor** to open the Serial Monitor.

2. The Arduino is now pinging the Google server every five seconds. If everything is working correctly, you should see a lot of `SUCCESS!` messages, like in my output:

```
Attempting to connect to WPA SSID: hodor
You're connected to the network
SSID: hodor
BSSID: 7C:FF:4D:22:C6:03
signal strength (RSSI): -56
Encryption Type: 4
IP address : 192.168.178.44
Subnet mask: 255.255.255.0
Gateway IP : 192.168.178.1
MAC address: 80:2A:3B:86:17:F0
Pinging www.google.com: SUCCESS! RTT = 30 ms
Pinging www.google.com: SUCCESS! RTT = 30 ms
Pinging www.google.com: SUCCESS! RTT = 30 ms
```

```
Pinging www.google.com: SUCCESS! RTT = 30 ms
Pinging www.google.com: SUCCESS! RTT = 20 ms
```

Great! We now know that your network username and password are correct and that your Arduino can access the internet. If you run into connectivity problems and you are not sure whether it is due to your network connection, you can just upload this sketch to find out whether the connection is still working and whether you should look somewhere else for the error.

Save this sketch as `ch4_network_check` in a folder for this book called `mqtt_arduino_book`.

Testing Arduino MQTT connectivity

Now, it's time to make sure MQTT works on your Arduino MKR WiFi 1010.

As of writing this, there is no official example of the MQTT-Arduino library for the Arduino MKR WiFi 1010. This is because the Arduino MKR WiFi 1010 uses a wireless module that is relatively new.

1. Create a new sketch, save it as `ch5_mqtt_test`, and paste the following code.

 Feel free to just copy and paste the code for this sketch. You can find it in this book's repository (`https://github.com/PacktPublishing/Hands-On-Internet-of-Things-with-MQTT`) in the folder.

2. First, there are the `include` statements and variables:

```
#include <WiFiNINA.h>
#include <MQTT.h>
const char WIFI_SSID[] = "YOUR_WIFI_NAME_HERE";
const char WIFI_PASS[] = "YOUR_WIFI_PASSWORD_HERE";
const char mqttServer[] = "broker.shiftr.io";
const int mqttServerPort = 1883;
const char key[] = "try"; // MQTT server username
const char secret[] = "try"; // MQTT server password
const char device[] = "YOUR_NAME-arduino"; // device identifier

int status = WL_IDLE_STATUS;
WiFiClient net;
MQTTClient client;

unsigned long lastMillis = 0;
```

3. Then, there is the `connect` function, which makes sure a Wi-Fi and MQTT connection are established. The first part establishes a Wi-Fi connection:

```
void connect() {
    Serial.print("checking wifi...");
    while ( status != WL_CONNECTED) {
        status = WiFi.begin(WIFI_SSID, WIFI_PASS);
        Serial.print(".");
        delay(1000);
    }
    Serial.println("\nconnected to WiFi!\n");
```

The second part of the `connect` function takes care of establishing the MQTT connection to the server:

```
    client.begin(mqttServer, mqttServerPort, net);

    Serial.println("connecting to broker...");
    while (!client.connect(device, key, secret)) {
        Serial.print(".");
        delay(1000);
    }
```

The last part of the connect function is executed once the Wi-Fi and MQTT connections are established:

```
    Serial.println("Connected to MQTT");
    client.onMessage(messageReceived);
    client.subscribe("/hello");
}
```

4. The `messageReceived` function is as follows:

```
void messageReceived(String &topic, String &payload) {
    Serial.println("incoming: " + topic + " - " + payload);
}
```

5. Then, there is `setup`, as well as the `loop` function:

```
void setup() {
    Serial.begin(115200);
    connect();
}

void loop() {
    client.loop();
    // delay(1000); // helps eventually
```

```
    if (!net.connected()) {
        connect();
    }

    if (millis() - lastMillis > 1000) {
        lastMillis = millis();
        client.publish("/hello", "world");
        lastMillis = millis();
    }
}
```

Let me explain what is happening here. First, we import the WiFiNINA network library, which we used in our first test and which is needed for the Arduino MKR WiFi 1010 to go online; then we import the MQTT library.

We then have to provide our network username and password (the same as in the network test before) and the MQTT server we want to connect to.

In our case, we are using the public credentials for shiftr.io (the username is try and the password is try), which you can use without signing up for an account. This means that anybody can read what you are sending to the MQTT server and anybody can send you MQTT messages as well, so this is not secure, but for prototyping purposes, it's absolutely okay.

The setup method is executed once the Arduino is powered on, when it is restarted or when new code has been uploaded. In this case, it opens a serial connection with Serial.begin(115200), so we can see whether everything is working fine or if there are any errors (in the Serial Monitor). Afterward, we call the connect method, which first tries to establish a network connection by using your network username and password and then connects to the MQTT server. The while loop makes sure it only leaves this function once a connection to the MQTT server has been established.

We define what should happen when there is an incoming MQTT message with the following line:

```
client.onMessage(messageReceived);
```

This tells the MQTT library to call the messageReceived function, which is defined once an MQTT message to one of the subscribed topics is available. Inside this function, we will, later on, add the functionality to control our servo motor.

After the setup is finished, the `loop` function is executed indefinitely. We first use a check at the beginning to make sure that we are still connected to the MQTT server:

```
if (!net.connected()) {
    connect();
}
```

If the Arduino was disconnected for some reason, it will reconnect by calling the `connect` function again.

The last part of the `loop` function publishes an MQTT message every second (in Arduino, time is specified in milliseconds, hence `1000`):

```
if (millis() - lastMillis > 1000) {
    lastMillis = millis();
    client.publish("/hello", "world");
    lastMillis = millis();
}
```

The `millis()` function is used to get the milliseconds that have passed since the Arduino was started, which is handy to use for timing tasks. In this case, we use it to detect whether the time passed since the code inside the `if` loop was executed is greater than 1,000 milliseconds—or 1 second. If so, we publish an MQTT message with the content, `world`, to the `/hello` channel.

The last line of the `setup` function, `client.subscribe("/hello");`, subscribes to the `/hello` topic, so every time an MQTT message is published to this channel, we will be notified about it:

1. Now save the sketch and upload it to your Arduino by pressing the **Upload** button.
2. To see whether everything is working as expected, open the Serial Monitor (**Tools | Serial Monitor**) and have a look at the output.

Please always make sure, when using the `Serial` library, that the baud rate is correctly set in the Arduino Serial Monitor. In our case, we use `115200` as the baud rate, as you can see in our `setup` function: `Serial.begin(115200);`. The baud rate set in the Arduino Serial Monitor must match this number.

If there are no errors, you should see something like this:

```
connected to WiFi!
connecting to broker...
Connected to MQTT
incoming: /hello - world
incoming: /hello - world
incoming: /hello - world
incoming: /hello - world
incoming: /hello - world
incoming: /hello - world
incoming: /hello - world
incoming: /hello - world
incoming: /hello - world
incoming: /hello - world
```

Because we are using a public namespace here and the /hello channel, which other people might use as well, let's pick a new one—say, /tims-channel, which others are less likely to use.

For this, you need to change two lines in the code:

1. First, we need to change which channel we subscribe to:

```
client.subscribe("/hello");
```

2. Change it to this:

```
client.subscribe("/tims-channel");
```

3. Now, we need to make sure to change where we send messages to as well:

```
client.publish("/hello", "world");
```

4. That needs to be changed to this:

```
client.publish("/tims-channel", "world");
```

The second parameter, world, can stay the same because it is just the topic we want to change.

If you upload the sketch and open the Serial Monitor again, you will see the output:

```
connected to WiFi!
connecting to broker...
Connected to MQTT
```

```
incoming: /tims-channel - world
incoming: /tims-channel - world
incoming: /tims-channel - world
incoming: /tims-channel - world
```

Okay, great! We are now able to subscribe to topics on the shiftr.io MQTT server and publish messages to the same channel. This means that any device in the world connected to the same MQTT server can send messages to the Arduino now and receive messages sent from the Arduino.

If you cannot wait until we do it together later on, feel free to try any of the MQTT apps for Android or iOS introduced in Chapter 3, *Getting Started with MQTT*. In these apps, you need to provide the following information to interact with your Arduino over the internet via MQTT:

- **MQTT server**: broker.shiftr.io
- **MQTT username**: try
- **MQTT password**: try
- **Channel to publish to**: /tims-channel
- **Channel to subscribe to**: /tims-channel

Feel free to use your own channel. Just make sure you change every occurrence of /tims-channel to your name, for example, /kathrins-channel.

If you have not done so, save this sketch as ch5_mqtt_test. This is an important sketch that you can later on reuse to make sure your Arduino MKR WiFi 1010 can connect to your Wi-Fi as well as the shiftr.io MQTT server.

If you have any problems so far, feel free to open an issue on this book's repository on GitHub (https://github.com/PacktPublishing/Hands-on-Internet-of-Things-with-MQTT) and I will help you to find a solution.

Testing the servo motor

Now that your network connection is working, as is the sending and receiving of MQTT messages, we need to do yet another mini-sketch to test whether the servo motor is working. Afterward, we will put the pieces of the puzzle together and create the code for our smart pet food dispenser.

To connect the servo motor, you need to connect the three cables of the servo to the Arduino. To connect the servo motor cable (with female pins) to the female Arduino pin headers on top, you can use the male-male jumper wire cables.

Connect the cables as follows:

- **Black (or brown)**: GND
- **Red**: VCC
- **Orange (or yellow/green/white)**: Pin 9

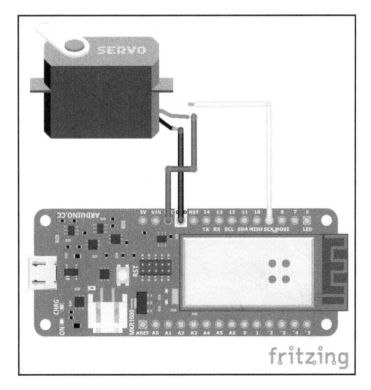

Only three connections are needed to connect the servo: VCC, ground, and pin 9(this image was created with Fritzing)
License : (https://creativecommons.org/licenses/by-sa/3.0/)

This is how it actually looks using the Tower Pro SG90 Servo Motor:

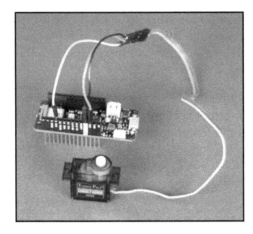

Servo motor connected via jumper wires to the Arduino

Normally, it would be preferable to start on a breadboard instead of doing direct connections from component to microcontroller board, but, in this case, we just have three cables, so we save a lot of space by doing the wiring as in the preceding photo. A connection via a breadboard would look like this:

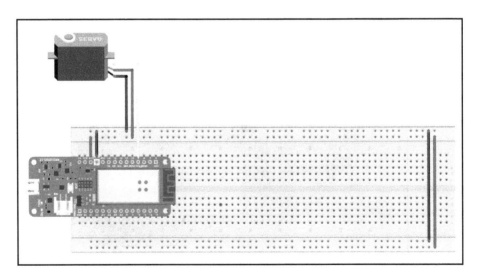

Alternative wiring with a breadboard (this image was created with Fritzing)
License : (https://creativecommons.org/licenses/by-sa/3.0/)

Now, open the Servo Sweep example (**File** | **Examples** | **Servo** | **Sweep**) included in the Arduino IDE.

Maybe you are wondering about the meanings of the colors now. Two colors have a special meaning when working with electronics. Red is used for the power line, and black is used for ground. These two are always needed to close the circuit. The third cable on the servo is the data line. Here the Arduino sends commands to the servo, telling it, for example, to go to position 120.

In the example code, we continuously send the command to update the servo position (with a different number), thereby making the servo rotate. We basically tell the servo to move one degree, move two degrees, move three degrees, and so on. In contrast to a DC motor, you cannot tell the servo motor to just spin; you always have to define to which position it should move.

Now upload the servo sweep example and the servo should start moving, in alternating directions.

Sometimes it happens that the code cannot be uploaded; in such a case, you should check whether the right board and port are selected in **Tools**.

Controlling the servo motor via the Serial Monitor

In the previous section, we verified the following:

- Whether the Arduino can connect to the internet
- Whether the Arduino can send and receive MQTT messages via shiftr.io
- Whether the motor is working

Now, let's create an interface to control the servo from the Serial Monitor. Create a new sketch and save it as ch5_01_servo_serial.

Delete the boilerplate code that is automatically added to new sketches (the empty `setup` and `loop` functions) and replace the contents of your editor with the following code:

```
#include <Servo.h>

Servo myservo;

void setup() {
  Serial.begin(9600);
```

```
    myservo.attach(9);
}

void loop() {
  while (Serial.available() > 0) {
    int inputValue = Serial.parseInt();
    if (inputValue == 1) {
      myservo.write(180);
    } else {
      myservo.write(0);
    }
  }
}
```

Let's go through the code to make sure you understand it. The first two lines load the `servo` library and create an instance of the `servo` class for us to use:

```
#include <Servo.h>
Servo myservo;
```

In the `setup` function, we do two things: we initialize the serial connection (so we can control the servo via serial messages) and we let the `servo` library know that we connected the servo to pin 9 of the Arduino:

```
void setup() {
    Serial.begin(9600);
    myservo.attach(9);
}
```

To initialize the serial connection, we use a baud rate of 9600. While this might seem like a cryptic number, you only need to know two things about it:

- When you exchange information between two devices via the `Serial` interface, both sides must use the same baud rate.
- Higher baud rates are needed for fast communication. In our case, 9600, which is relatively slow, is enough.

We want to be able to move the servo to one of two positions, so it can be used to open and close our dispenser, which we are about to build.

As such, we need to define two commands that move the servo to either the open position or the closed one. To keep it simple, let's use the number 1 to open it, and the number 0 to close it.

Let's have a look at the `loop` function:

```
void loop() {
  while (Serial.available() > 0) {
    int inputValue = Serial.parseInt();
    if (inputValue == 1) {
      myservo.write(180);
    } else {
      myservo.write(0);
    }
  }
}
```

In the `loop` function, which is called repeatedly after the `setup` function, we check whether any characters are waiting for us in the serial buffer and, if so, we read them as a whole number (an integer). We then check whether the number we just read in from serial is `1`. If it is, we tell the servo to move to position `180`. If it is any other character than `1`, the servo should move to position `0`.

Now, re-upload the code and open the Serial Monitor (**Tools | Serial Monitor**).

Now, enter `1` into the input field; then hit the *Enter* key to send the command. Try also to enter `0`. The servo moves—hooray!

We have all of the code together now to actually build the dispenser and control it via serial messages.

Building the smart pet food dispenser

Let's gather all of the materials we need to make the dispenser using the servo motor. Once we are finished with it, we will add MQTT functionality, so you can control it via MQTT (for example, via a smartphone app).

We will need the following:

- A plastic bottle with a big opening
- Scissors
- Thick paper or thin cardboard
- Hot glue or super glue
- Tape

While building the smart pet food dispenser, for some steps, you will need to move the servo motor by hand. Try to keep manual rotation to a minimum. Excessively rotating it by hand might eventually destroy it.

Follow the following steps to build the dispenser:

1. Cut the bottle in half using the scissors. We will use the top part for our food dispenser:

Servo motor attached to the half bottle with tape

2. To simulate the dispenser functionality, hold the bottle upside down and fill in the food you want to dispense (for example, peanuts). Does it fit through the bottle opening? You can simulate the behavior of the automatic pet food dispenser by manually opening and closing the bottle opening using the piece of paper or cardboard that you move with your hands. If your food gets stuck, you might need a bottle with a bigger opening. If the food fits through the opening well, continue with the next step.

3. Your servo motor probably came with different caps to put on. Pick the shortest one and put it on the servo motor.

4. Attach the servo motor body to the plastic bottle with tape. Make sure that the servo is still able to move freely.

If you need a longer cable between the Arduino and the servo, you can chain together flexible breadboard cables.

5. Find out how to best put the cap on, so that, in one position, it points toward the opening of the bottle and, in the other position, 90° away from the bottle opening.

In a later step, we will attach the cardboard to the servo cap; for now, just imagine it is already attached to the servo cap.

6. Find out the perfect value for the open position of the servo. In the open position, the servo cap should point away from the bottle. Send 1 via the Serial Monitor to the Arduino. Does it point away from the servo already? Then, you don't have to do anything in this step. If it does not point away from the bottle opening, you need to change the value that controls the angle of the servo in the open position:

```
myservo.write(180)
```

Change it to another value between 0 and 180, upload your code, then send 1 via the Serial Monitor again. Repeat this until you find the perfect value so that, when you send 1 via Serial Monitor, the cap points away from the opening.

7. Similar to the previous step, we now need to find a good value for the closing position. In the closing position, the cap should point toward the opening of the bottle. Send 0 via the Serial Monitor to move the servo to the closing position. Change the value of the following line accordingly:

```
myservo.write(0)
```

Change the number to a value between 0 and 180; then re-upload the code and send 0 via the Serial Monitor to close the dispenser. Once you find the perfect setting, sending 0 should close the dispenser and 1 should open it.

Don't worry if it doesn't look good; we will attach something to hide it later on. The inner part just needs to be functional, not appealing. In general, when building prototypes, it does not matter too much how the inner part of the prototype looks. Nobody will see it. If it works and it looks okay from the outside, it is absolutely fine.

8. Cut a piece of cardboard as in the following photograph. It should be a little bit bigger than the opening of the bottle:

Servo cap glued onto a small piece of cardboard

9. Attach the bottle cap to the piece of cardboard with hot glue or super glue. Make sure it still fits on the servo and still covers the bottle opening.

10. Find a good position to put the newly created cap on. Does it open and close well when sitting on the servo motor? Try turning it slowly to find out.

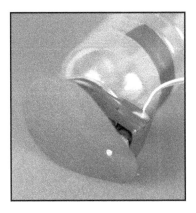

Servo cap screwed onto the servo

11. To prevent our cap from falling down when being used, we can use a tiny screw (not included in the servo motor package) to tightly connect the servo cap to the servo shaft. If you have a collection of leftover screws, have a look at whether you can find a screw that fits into the servo shaft. You can also do this later if you notice the cap falling down all of the time.

Another solution to this problem is adding a little bit of super glue between the servo shaft and the servo cap. You just need to make sure that it is just one or two drops of super glue that you use. Using too much glue might lead to the glue flowing into the motor and the motor being blocked. Finding a tiny screw is recommended over this method.

Optimizing the dispenser code

Now, let's get back to coding. You probably tried out how the dispenser works with real material in it, controlled by sending commands as 1 and 0 over the Serial Monitor. Sending two commands to manually open and close the opening is not super useful.

Let's modify our code so that the servo automatically goes back to the closed position after x milliseconds. If you use a very fine material (such as sugar), this might be something like 150 milliseconds. For grainier material (dry cat food, for example), something in the seconds might be good. Remember, we always need to specify the time in milliseconds, so 2 seconds would be 2,000 milliseconds.

Save the existing sketch as a copy with the name ch5_02_servo_serial_auto_close.

To automatically close the cap shortly after it is opened, we need to add a few new variables. Follow these steps:

1. Add the following code directly after the Servo myservo; line:

```
unsigned long lastTimeOpenend = 0;
bool isOpen = false;
int OPENING_TIME = 700;
```

lastTimeOpenend will hold the time in milliseconds when the dispenser was last opened. Time is stored in milliseconds, so this number can become quite big.

We need to use the unsigned long variable type for this, which can store bigger (positive) whole numbers than int, which we would use for most use cases. You can read more about it on the Arduino reference (https://www.arduino.cc/reference/en/language/functions/time/millis/).

The next variable, isOpen, will be set to true when we receive an open command and open the dispenser. We also define the time in milliseconds that the dispenser should stay open for, OPENING_TIME. In my case, 700 is a good value. Good values for you might be in the range from 50 to 4000. See what works best for you.

We also need to edit our `loop` function to include the auto-close code.

2. Replace your `loop` function with the following code:

```
void loop() {
    while (Serial.available() > 0) {
        int inputValue = Serial.parseInt();
        if (inputValue == 1) {
            lastTimeOpenend = millis();
            isOpen = true;
            myservo.write(90);
        }
    }
    if (isOpen) {
        if (millis() > lastTimeOpenend + OPENING_TIME) {
            myservo.write(0);
            isOpen = false;
        }
    }
}
```

In the previous section, you fine-tuned the values for the angles that are passed to the `myservo.write` calls yourself, as this depends on the type of servo motor you are using and how you attached it to the plastic bottle and cap. Make sure to not lose these two values when you enter the preceding code.

In the `if` clause, we added two lines:

```
if (inputValue == 1) {
    lastTimeOpenend = millis();
    isOpen = true;
    myservo.write(90);
}
```

`lastTimeOpenend = millis();` stores the current time in milliseconds since the Arduino started in the `lastTimeOpenend` variable.

Also, `isOpen` is set to `true`.

We also added another `if` clause:

```
if (isOpen) {
    if (millis() > lastTimeOpenend + OPENING_TIME) {
        myservo.write(0);
        isOpen = false;
    }
}
```

First, we check whether `isOpen` is `true`, and then we do another check to find out whether it is time to close the dispenser. This pattern is used very often to run specific code after a certain amount of time and you will very likely need it again. When the current time is greater than the time we opened the dispenser and `OPENING_TIME` milliseconds have passed, it is time to close the dispenser again. We send the new command to the servo to move to the closing position and, finally, set our `isOpen` variable to `false`.

When you reupload the sketch and send a `1` character, you should now see the dispenser opening for a short time and then automatically closing again.

Improving the visual appearance

Let's improve the visual appearance of our food dispenser:

1. Wrap a piece of paper or cardboard around the dispenser. This way, we can hide the plastic and servo:

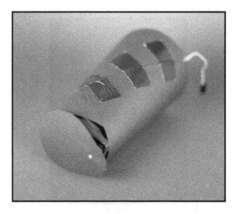

Cardboard sleeve around the plastic bottle, fixed using tape

2. You can use DIN A4 cardboard to build the sleeve. Wrap it around, fix it using tape, and then cut off the top part (next to where the bottle ends) to make it evenly round.

3. To hide the inner parts even more, we can build another cylinder-like part around the bottom part of our dispenser (near the cap):

Another piece of cardboard is wrapped around the bottom part to form a cylinder and hide the servo motor

4. Before fixing the cylinder-shaped cardboard, try moving the cap to see whether it still opens and closes properly. When you find a good position, fix it with tape:

The second cylinder-shaped cardboard hides the servo motor

Depending on the type of bottle you have, you might find a better solution for how to hide the ugly parts.

Let's move on and connect the dispenser to the internet.

Making the dispenser controllable via MQTT

Now let's add MQTT functionality. What we want to achieve is the following.

The Arduino should subscribe to the `/yourname/feeder/feed` topic. The namespace consists of three parts: `yourname`, which you can freely pick, should unify all of your MQTT experiments; `feeder`, which is the name of your device (you could also give it a longer name, such as `food-dispenser`); and `feed`, which is a command we use to release one portion of food. We could also add other commands here. One command that would be good to add is `status`, to which the device sends a message when it connects or disconnects to the internet and MQTT server. Its namespace would look like this: `/yourname/feeder/status`. Feel free to add this functionality later on.

To make our feeder accessible via MQTT, we need to combine our current sketch with the MQTT test code we ran earlier. Save our current code as `ch5_03_servo_mqtt`.

We need to do a lot of copying and pasting now. If you get lost somewhere, it is no problem. I will provide the complete code later on—but you should try to actively follow along anyway.

Being able to combine multiple sketches into one basically gives you superpowers. There are so many examples and code snippets out there in the world that do a certain thing and showcase how to do it. If you know how to combine them, you can achieve a lot without needing to code much yourself. And yes, that's a good thing. With each of my Arduino projects, I start the same way. I look for examples that do what I need, sometimes trying out a lot of examples until I find what I need; then I combine them and add some extra code. Believe me, you don't need to write much code to make interesting, beautiful, and useful smart things. You should have a basic understanding of what the code does, though. Also, not being able to code will limit you, but you can get pretty far if you can do one thing well: combine existing sketches.

To add MQTT functionality to your smart pet food dispenser, follow these steps:

1. Open the `ch4_mqtt_test_public` MQTT test sketch and view it side by side with our current sketch, `ch5_03_servo_mqtt`.

   We plan to integrate all the bits and pieces we need from `ch4_mqtt_test_public` into our current sketch.

When integrating one sketch into another, there are a few things that you have to look out for. Header includes (such as `#include <MQTT.h>`) and variable declarations from the top of the sketch can be copied as they are. You just have to look out for double-including them. If, for example, our current sketch already had the line, `#include <MQTT.h>`, so you would not need it twice.

2. Copy the include statements from the MQTT test sketch, as well as all of the variable declarations (`const char WIFI_SSID[]` and so on) into our current sketch.

It is a good practice to copy the includes separately from the variable declarations to put them at the beginning of the sketch. You should do the same for the variable declarations. The first part of your sketch should look like this now:

```
#include <Servo.h>
#include <WiFiNINA.h>
#include <MQTT.h>

const char WIFI_SSID[] = "YOUR_NETWORK_NAME";
const char WIFI_PASS[] = "YOUR_PASSWORD";
const char mqttServer[] = "broker.shiftr.io";
const int mqttServerPort = 1883;
const char key[] = "try";
const char secret[] = "try";
const char device[] = "hellomqtt"; // broker device identifier

Servo myservo;
unsigned long lastTimeOpenend = 0;
bool isOpen = false;
int OPENING_TIME = 700;
int status = WL_IDLE_STATUS;
WiFiClient net;
MQTTClient client;
unsigned long lastMillis = 0;
```

Now, we need to integrate the code from `setup`. There can only be one `setup` function for each sketch, as well as one `loop` function.

3. Copy the lines from inside the MQTT `setup` function into the `setup` function of our current sketch. These are the two lines:

```
Serial.begin(115200);
connect();
```

4. Paste them at the end of our current `setup` function:

```
void setup() {
    Serial.begin(9600);
    myservo.attach(9);
    Serial.begin(115200);
    connect();
}
```

Whoops! Seems like we have two `Serial.begin` lines now—that needs to be fixed. Do you remember the parameter of the `Serial.begin` function, the baud rate? It defines how fast characters are transmitted via serial between the sender and the receiver. As we don't need to send so much via `Serial` (only some status information), let's keep the `Serial.begin(9600)` line and get rid of the other one, `Serial.begin(115200)`.

5. Remove the `Serial.begin(115200)` line. Your `setup` function should look like this now:

```
void setup() {
    Serial.begin(9600);
    myservo.attach(9);
    connect();
}
```

After initializing the `Servo` library, we call `connect`. We haven't copied that one over yet from the MQTT test sketch. Let's do so now. We will also need the `messageReceived` function, which gets called whenever there is an incoming MQTT message.

6. Copy both the `connect` and `messageReceived` functions before the `setup` function in our current sketch.

Now, the only thing left to integrate is the code inside the `loop` function. As we can only have one `loop` function, we cannot just copy the whole function; we have to integrate the contents of the `loop` function.

7. Copy everything inside the `loop` function of the MQTT test sketch.
8. Paste it at the very beginning of the `loop` function of our main sketch. Your whole code should look like this now.

First, there are the `include` statements and variable definitions:

```
#include <Servo.h>
#include <WiFiNINA.h>
#include <MQTT.h>

const char WIFI_SSID[] = "YOUR_NETWORK_NAME";
const char WIFI_PASS[] = "YOUR_NETWORK_PASSWORD";
const char mqttServer[] = "broker.shiftr.io";
const int mqttServerPort = 1883;
const char key[] = "try";
const char secret[] = "try";
const char device[] = "hellomqtt"; // broker device identifier

Servo myservo;
unsigned long lastTimeOpenend = 0;
bool isOpen = false;
int OPENING_TIME = 700;
int status = WL_IDLE_STATUS;
WiFiClient net;
MQTTClient client;
unsigned long lastMillis = 0;
```

Then, there is the `connect` function. The first part takes care of establishing the Wi-Fi and MQTT connection:

```
void connect() {
    Serial.print("checking wifi...");
    while ( status != WL_CONNECTED) {
        status = WiFi.begin(WIFI_SSID, WIFI_PASS);
        Serial.print(".");
        delay(1000);
    }
    Serial.println("\nconnected to WiFi!\n");

    client.begin(mqttServer, mqttServerPort, net);

    Serial.println("connecting to broker...");
    while (!client.connect(device, key, secret)) {
        Serial.print(".");
        delay(1000);
    }
```

The second part of the `connect` function logs the success message to `Serial`, defines the function that is called once an MQTT message arrives and subscribes to an MQTT channel:

```
Serial.println("Connected to MQTT");

client.onMessage(messageReceived);

client.subscribe("/tims-channel");
// client.unsubscribe("/hello");
}
```

`connect` should be followed by the `messageReceived` and `setup` functions:

```
void messageReceived(String &topic, String &payload) {
    Serial.println("incoming: " + topic + " - " + payload);
}

void setup() {
    Serial.begin(9600);
    myservo.attach(9);
    connect();
}
```

Then, there is the `loop` function. The first part takes care of keeping the Arduino connected to the Wi-Fi and MQTT server and publishes a message every second:

```
void loop() {
    client.loop();

    if (!net.connected()) {
        connect();
    }

    if (millis() - lastMillis > 1000) {
        lastMillis = millis();
        client.publish("/tims-channel", "world");
        lastMillis = millis();
    }
```

The second part of the `loop` function handles incoming serial characters and takes care of automatically closing the dispenser after it is opened:

```
while (Serial.available() > 0) {
    int inputValue = Serial.parseInt();
    if (inputValue == 1) {
        lastTimeOpenend = millis();
        isOpen = true;
```

```
            myservo.write(90);
        }
    }
    if (isOpen) {
        if (millis() > lastTimeOpenend + OPENING_TIME) {
            myservo.write(0);
            isOpen = false;
        }
    }
}
```

9. Save your sketch and click the verify icon to see whether there were any errors. Hopefully, everything compiled well and you will see the `done compiling` message at the bottom of the window.

   We need to make some tiny modifications now to achieve what we want. The MQTT test sketch publishes a message every two milliseconds. We don't really need this code, but it might be helpful for debugging, later on, to see that our Arduino is online and connected to the MQTT server.

10. Find the `if (millis() - lastMillis > 1000) {` line, which we use to check whether one second has passed and it is time to republish the MQTT message. Change it to 10 seconds by appending `0`, so that `1000` becomes `10000`.

11. We also want a more meaningful message. Change the following line:

    ```
    client.publish("/tims-channel", "world");
    ```

    Change it to this:

    ```
    client.publish("/tims-channel/feeder/ping", "feeder is
    online");
    ```

    You probably want to change it to reflect your name, instead of mine.

12. Re-upload the code. We are now sending a message to the `/tims-channel/feeder/ping` channel every 10 seconds. Feel free to change the `feeder is online` string to something else.

Now comes the important part: we want to subscribe to `/tims-channel/feeder/feed`. If we receive a message to this channel, we want to open the feeder for a short time to let food out and close it automatically afterward.

There is not much left to do for this. If you look at the `setup` function, you will see that we already subscribed to `/tims-channel` by calling `client.subscribe("/tims-channel")`. We only need to make a tiny change to also receive `/tims-channel/feeder/feed`.

In `Chapter 3`, *Getting Started with MQTT,* we talked about wildcards. The multi-level wildcard, `#`, can be used to subscribe to all sub-topics.

13. Change the line that starts with `client.subscribe` to the following:

```
client.subscribe("/tims-channel/feeder/#");
```

This subscribes to `/tims-channel/feeder/feed` and possibly other sub-channels, which we might later add.

Now we need to add our main logic. Every time we send an MQTT message to `/tims-channel/feeder/feed`, the feeder should open to release some food, and shortly after, close again. The function that is called when there is a new MQTT message waiting for us is `messageReceived`. Here we need to add a check for whether the topic of the incoming message is `/tims-channel/feeder/feed` and not `/tims-channel/feeder/ping` or something else such as `/tims-channel/feeder/just-a-test`.

14. Replace your `messageReceived` function with the following:

```
void messageReceived(String &topic, String &payload) {
    Serial.println("incoming: " + topic + " - " + payload);
    if (topic == "/tims-channel/feeder/feed") {
        Serial.println("Command to feed received");
    }
}
```

We also need to copy some code from our `loop` function into the `messageReceived` function to actually move the servo.

15. Scroll down to our `while` loop:

```
while (Serial.available() > 0) {
    int inputValue = Serial.parseInt();
    if (inputValue == 1) {
        lastTimeOpenend = millis();
        isOpen = true;
        myservo.write(90);
    }
}
```

16. Copy the three lines inside the `if` statement, that is, everything between `if (inputValue == 1) {` and the closing bracket `}`. This is the code to move the servo that is executed when we send the `1` character via the Serial Monitor:

```
lastTimeOpenend = millis();
isOpen = true;
myservo.write(90);
```

17. Paste it inside our newly created `if` clause in the `messageReceived` function, so it looks like this:

```
void messageReceived(String &topic, String &payload) {
    Serial.println("incoming: " + topic + " - " + payload);
    if (topic == "/tims-channel/feeder/feed") {
        Serial.println("Command to feed received");
        lastTimeOpenend = millis();
        isOpen = true;
        myservo.write(90);
    }
}
```

> While, mostly, it is better to create a function and extract the code that is used in multiple places into it and then call the function, for simplicity, we will just copy the code as we did previously.

We are all set now. You should be able to send an MQTT message to this channel and your smart food dispenser should open and close shortly after. If you had any problems following along, you can find the complete code for this chapter in this book's repository (`https://github.com/PacktPublishing/Hands-on-Internet-of-Things-with-MQTT`) in the `ch5/arduino/ch5_03_servo_mqtt/` folder.

Sending commands to the dispenser via MQTT

Let's use the Mosquitto command-line tool to send the MQTT message to the MQTT server (shiftr.io), which will then forward it to our Arduino. Open a Terminal/PowerShell and run the following code:

```
mosquitto_pub -t "/tims-channel/feeder/feed" -m "Feeeed" -h
broker.shiftr.io -u "try" -P "try"
```

We use the `mosquitto_pub` command to send the `"Feeeed"` message to the `/tims-channel/feeder/feed` channel using the `broker.shiftr.io` MQTT server with the `try` username and `try` password. At this point, we will ignore the message, in this case `Feeeed`, in our Arduino code and just check whether the topic matches `/tims-channel/feeder/feed`.

When you run the code, your feeder should open up and close again. Hooray!

 You can send the command to open the dispenser from any of the apps we discussed in the app overview in Chapter 3, *Getting Started with MQTT*. In Chapter 7, *Build a Smart Productivity Cube, Part 1*, and Chapter 8, *Build a Smart Productivity Cube, Part 2*, you will learn how to make use of MQTT apps to interact with your prototypes.

As you can see, the finished food dispenser serves peanuts. Two holes were added to hang the dispenser using a piece of string:

The finished peanut dispenser prototype: the left is closed and the right is opened

You can also open the Serial Monitor to see the incoming messages (we also receive the ping message sent from the Arduino itself, because we subscribed to all messages sent to `/tims-channel/feeder` with the # wildcard).

You are probably sitting at a desk right now with your computer and the Arduino next to each other. Entering one command in the Terminal and having the servo motor next to your computer move might seem like a small thing, but thanks to MQTT, they are completely decoupled now. You could disconnect the Arduino and plug it into a USB wall charger, then leave your house and control the Arduino from anywhere in the world where you have internet access, either via your computer or your smartphone.

 Please keep in mind that this is a prototype, and therefore, you should not bet the life of your pet on it working correctly. The aim is to try it out while you are at home.

When I finished my first project using MQTT and realized that I could easily connect any prototype to the internet and control it from my smartphone, I felt really empowered. I hope you feel the same way and that you enjoyed the first project!

Summary

In this chapter, you learned how to control a servo motor connected to an Arduino MKR WiFi 1010 via MQTT. Using everyday household items such as a plastic bottle, we created a practical smart device—a smart (pet) food dispenser that you can control from anywhere in the world using an MQTT client (for example, an MQTT app for Android or iOS).

In Chapter 7, *Build a Smart Productivity Cube, Part 1*, and Chapter 8, *Build a Smart Productivity Cube, Part 2*, we will use an MQTT client for iOS and Android to interact with our prototypes. You can, later on, combine what you have learned there to control your automatic pet food dispenser with your smartphone.

The knowledge you have gained in this chapter allows you to create all sorts of smart objects that require a servo motor. If you are looking for inspiration, have a look at the Arduino Project Hub (`https://create.arduino.cc/projecthub`) and search for `servo`.

We were able to create the smart (pet) food dispenser with very little custom-written code, just by combining existing examples. This technique is very powerful because it allows you to create complex applications by combining existing examples or code snippets like the pieces of a puzzle, even without being an expert programmer.

In Chapter 6, *Building a Smart E-Ink To-Do List*, we will build a smart e-ink to-do list using the same technique: finding an example that fits our use case, making it work, and building upon it. The smart e-ink to-do list uses an e-ink display and can display textual information, such as "don't forget to take out the trash."

Questions

1. What does the # character in the topic name do?
2. Is using the /your-name/feeder topic really private/secure?
3. What can you do if shiftr.io stops working?
4. What possibilities do you have to control your smart pet food dispenser?
5. Why is combining existing examples/code snippets often preferable to writing everything from scratch?
6. Do you need to be a great coder to build functional useful prototypes?
7. Should you bet the life of your pet on this prototype working?
8. What else can you build with a servo motor that can be controlled from anywhere?

Building a Smart E-Ink To-Do List

6

Our use of MQTT in Chapter 5, *Building Your Own Automatic Pet Food Dispenser*, was very simplistic. We were only sending (empty) messages to a channel the Arduino was subscribed to. In this project, we will make use of the payload, which can be used to append text (or binary data) to an MQTT message.

Our mission is to build a smart e-ink display that can be used to remind yourself of tasks such as taking out the trash or buying milk. You can hang it next to your door so you can always see what needs to be done when you leave your flat or house. E-ink displays are great for displaying text or images that do not change often. You have probably heard of eBook readers before. They use e-ink displays as well. E-ink displays are very energy efficient and only consume energy when the content is being updated. Once new content is written to the display, it does not take up any more energy until new content is sent to it.

Since updating the display to show new text or images takes much longer than with a regular display, you would definitely not use it to watch a video or play games. Instead, such displays will be used to display text and images, which are not updated for a couple of minutes or hours.

After completing this project, you will be able to use an internet-connected e-ink display in any of your next projects. There are many use-cases specifying what you can do with an internet-connected e-ink display and how you can enhance a project using it. Using a web service such as IFTTT (https://ifttt.com), you can connect your display to a website such as Facebook or Twitter to always show the last message you received, display the weather, or the most recent news headlines.

The following topics will be covered in this chapter:

- Connecting the e-paper module/running the example
- Simplifying the e-paper example
- Modifying the e-paper example
- Making your e-paper device accessible via serial
- Preparing the MQTT integration
- Making your e-paper device accessible with MQTT
- Sending messages via MQTT
- Enhancements and building a case

Technical requirements

You will need the following components to build the smart e-ink to-do list:

- Arduino MKR WiFi 1010: `https://store.arduino.cc/usa/mkr-wifi-1010`
- Waveshare 4.2-inch e-ink display module, three-color (black, white, red), 400x300: `https://www.waveshare.com/4.2inch-e-paper-module-b.htm`
- Micro USB cable: `https://www.sparkfun.com/products/10215`

The following tools and libraries, which we installed together in `Chapter 4`, *Setting Up a Lab Environment*, are required:

- Arduino IDE
- Arduino MKR WiFi 1010 drivers
- WiFiNINA Arduino library
- MQTT Arduino library

Please download the source code from the book's repository (`https://github.com/PacktPublishing/Hands-on-Internet-of-Things-with-MQTT`). You will find all relevant sketches for this chapter in the `ch6` folder.

Check out the following video to see the Code in Action:
`http://bit.ly/2oSjufZ`

By now, I am assuming that you have completed the Arduino MKR WiFi 1010 setup, and successfully ran the MQTT example introduced in `Chapter 5`, *Building Your Own Automatic Pet Food Dispenser*. In case you run into any issues with connectivity, you should go back to these examples to make sure the basic network/MQTT code is working.

To make use of the Waveshare 4.2-inch e-ink display module, we need to install a library for it. There are various third-party libraries around, and it is not easy to pick the right one. In most cases, running a Google Search for the name of the electronic component you want to use together with the term Arduino will bring up tutorials or guides on how to connect the component to your Arduino. In this case, it was a bit tricky.

For e-ink displays, in this case a model from the company Waveshare, there are various libraries to choose from, and so far, only one seems to be compatible with the Arduino MKR WiFi 1010. The documentation of the libraries didn't help me to choose the right one, so I had to ask for help in the Arduino Forum (https://forum.arduino.cc). Luckily, the library creator, ZinggJM, was kind enough to tell me about his library GxEPD2, which is compatible with the Arduino MKR WiFi 1010.

Even if you have spent a lot of time with the Arduino IDE and know your way around, there will be many times when asking for help is the only sane thing to do. The users in the Arduino forum have been very helpful so far, and I cannot recommend enough creating an account there and asking for help, and after you have learned a thing or two, you can give back by helping others as well.

But let's move on. To make use of the e-ink display, we need to install the **GxEPD2** library via the Arduino **Library Manager**.

Open **Tools** | **Manage Libraries**, search for GxEPD2, and install the latest 1.x.x release (x standing here for any number). I briefly explained the scheme behind this numbering (called semantic versioning) in Chapter 4, *Setting Up a Lab Environment*:

The GxEPD2 library in the Arduino Library Manager

Now, follow these steps to get started:

1. The library we just installed depends on another library: the **Adafruit GFX Library**. In the Arduino Library Manager, search for **Adafruit GFX Library** and install it as well:

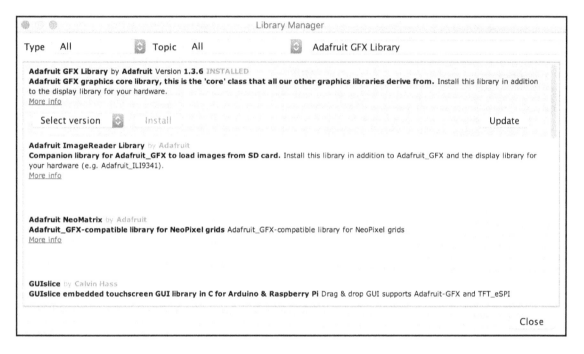

The Adafruit GFX Library in the Arduino Library Manager

2. Once the library is installed successfully, restart Arduino, reopen it, and you should find a new category, **GxEPD2**, when you navigate to **File** | **Examples**:

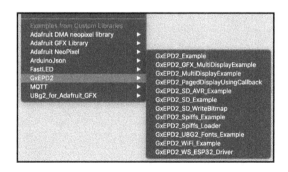

The GxEPD2 examples for the e-ink display

3. Open `GxEPD2_Example`:

```
● ● ●                         GxEPD2_Example | Arduino 1.8.8

 ✓ ➔ ▤ ⬆ ⬇                                                        🔎

 GxEPD2_Example   GxEPD2_boards_added.h                              ▾

 1 // Display Library example for SPI e-paper panels from Dalian Good Display and boards from Waveshare.
 2 // Requires HW SPI and Adafruit_GFX. Caution: these e-papers require 3.3V supply AND data lines!
 3 //
 4 // Display Library based on Demo Example from Good Display: http://www.e-paper-display.com/download_li
 5 //
 6 // Author: Jean-Marc Zingg
 7 //
 8 // Version: see library.properties
 9 //
10 // Library: https://github.com/ZinggJM/GxEPD2
11
12 // Supporting Arduino Forum Topics:
13 // Waveshare e-paper displays with SPI: http://forum.arduino.cc/index.php?topic=487007.0
14 // Good Dispay ePaper for Arduino: https://forum.arduino.cc/index.php?topic=436411.0
15
16 // mapping suggestion from Waveshare SPI e-Paper to Wemos D1 mini
17 // BUSY -> D2, RST -> D4, DC -> D3, CS -> D8, CLK -> D5, DIN -> D7, GND -> GND, 3.3V -> 3.3V
18
19 // mapping suggestion from Waveshare SPI e-Paper to generic ESP8266

                                              Arduino MKR WiFi 1010 on /dev/cu.usbmodem14101
```

GxEPD2_Example in Arduino

Next, we will connect the e-paper module and run the example.

Connecting the e-paper module/running the example

The next step is to connect the e-paper display to the Arduino MKR WiFi 1010. By looking at the connectors of the e-paper panel and comparing it to the available ports of the Arduino MKR WiFi 1010, we will see that the following port names match:

- GND
- VCC (3.3V)

Ground (GND) and **power (VCC)** are needed for every module, so no surprises here. For the other ports, we have to find out how to make it work.

My first step here would be to run a Google Search to find out if somebody posted a tutorial for the module we want to use with our microcontroller. This would mean searching for the Arduino MKR WiFi 1010 Waveshare 4.2 e-paper module.

I could not find any results (probably because Arduino MKR WiFi 1010 was very new when writing the book). To find out which ports to use, the best bet is to search for the official documentation from the manufacturer.

Searching for Waveshare 4.2 e-paper module brings us to the website of the manufacturer that supplies a user manual, a datasheet, and a demo code. Looking at the user manual (`https://www.waveshare.com/w/upload/2/20/4.2inch-e-paper-module-user-manual-en.pdf`) offers some valuable insights:

- The e-paper panel should be used with a 3.3V power supply, not 5V. The Arduino MKR WiFi 1010 runs on 3.3V, so we are safe here. It also has a 5V port, but we will not use it.
- The display panel has to be refreshed in certain intervals, otherwise, it might get damaged.
- It uses an **SPI interface** (**serial peripheral interface**), which is a very common interface for hardware modules.

Looking at the documentation further, we get some more hints on the wiring, as follows:

- **VCC**: 3.3V/5V
- **GND**: GND
- **DIN**: SPI MOSI pin
- **CLK**: SPI SCK pin
- **CS**: SPI chip selection, low active
- **DC**: Data/command selection (high: data; low: command)
- **RST**: Reset (low active)
- **BUSY**: Busy (low active)

We already know that VCC on the panel needs to be connected to VCC on the Arduino, and GND on the panel needs to be connected to GND on the Arduino. The rest gives us some more insight. DIN needs to be connected to SPI MOSI. **MOSI** stands for **Master Out Slave In**. Master, in this case, is the microcontroller, and slave is the e-paper panel. The SPI interface can be used to have bi-directional communication. MOSI sends data from the microcontroller to the display, and in this case, information on what to display.
By looking at the Arduino MKR WiFi 1010, we see a MOSI pin, so we know DIN has to be connected to MOSI (pin 8).

The next pin we need to find out how to connect is the **CLK** pin (which stands for **clock**). The panel documentation lets us know that it needs to be connected to the SPI SCK pin (**SCK** stands for **Serial Clock**). By looking at the Arduino MKR WiFi 1010, we also find an SCK pin (pin 9). Great, so we know where VCC, GND, DIN, and CLK have to be connected to. All the other pins can be defined in the code, as we will see in a bit.

I could just tell you which pins to connect; that would be easier, but I don't want you to just make the e-panel work, but also to learn how to find out how to connect it, so with the next module you buy, it will be easier for you to make it work yourself.

Let's prepare the code to run the e-paper example and find the missing pins:

1. Go back to the e-paper Arduino example we opened up earlier—GxEPD2_Example. By scrolling through once, you will notice that there is a lot of code. Really a lot. Don't worry. We don't need all of it to use the e-paper module. The example includes a lot of code for different microcontrollers, different e-paper modules, and use cases.

   To use it for the Arduino MKR WiFi 1010, we have to uncomment one line of code, which is specific to SAMD microcontrollers (which the Arduino MKR WiFi 1010 uses).

2. Navigate to the GxEPD2_boards_added.h tab, scroll down a bit, and you will find the following line:

   ```
   #if defined(ARDUINO_ARCH_SAMD)
   ```

   Everything between this line and #endif, which is a few lines below, will only be executed when the connected board uses a SAMD microcontroller.

   The whole block is relevant to us:

   ```
   #if defined(ARDUINO_ARCH_SAMD)
    #define MAX_DISPAY_BUFFER_SIZE 15000ul // ~15k is a good
   compromise
    #define MAX_HEIGHT(EPD) (EPD::HEIGHT <= MAX_DISPAY_BUFFER_SIZE
   / (EPD::WIDTH / 8) ? EPD::HEIGHT : MAX_DISPAY_BUFFER_SIZE /
   (EPD::WIDTH / 8))
    // select one and adapt to your mapping
    //GxEPD2_BW<GxEPD2_154, MAX_HEIGHT(GxEPD2_154)>
   display(GxEPD2_154(/*CS=4*/ 4, /*DC=*/ 7, /*RST=*/ 6, /*BUSY=*/
   5));
    //GxEPD2_BW<GxEPD2_213, MAX_HEIGHT(GxEPD2_213)>
   display(GxEPD2_213(/*CS=4*/ 4, /*DC=*/ 7, /*RST=*/ 6, /*BUSY=*/
   5));
   ```

```
//GxEPD2_BW<GxEPD2_213_flex, MAX_HEIGHT(GxEPD2_213_flex)>
display(GxEPD2_213_flex(/*CS=*/ 4, /*DC=*/ 7, /*RST=*/ 6,
/*BUSY=*/ 5)); // GDEW0213I5F
//GxEPD2_BW<GxEPD2_290, MAX_HEIGHT(GxEPD2_290)>
display(GxEPD2_290(/*CS=*/ 4, /*DC=*/ 7, /*RST=*/ 6, /*BUSY=*/
5));
//GxEPD2_BW<GxEPD2_290_T5, MAX_HEIGHT(GxEPD2_290_T5)>
display(GxEPD2_290_T5(/*CS=*/ 4, /*DC=*/ 7, /*RST=*/ 6,
/*BUSY=*/ 5)); // GDEW029T5
//GxEPD2_BW<GxEPD2_270, MAX_HEIGHT(GxEPD2_270)>
display(GxEPD2_270(/*CS=*/ 4, /*DC=*/ 7, /*RST=*/ 6, /*BUSY=*/
5));
//GxEPD2_BW<GxEPD2_420, MAX_HEIGHT(GxEPD2_420)>
display(GxEPD2_420(/*CS=*/ 4, /*DC=*/ 7, /*RST=*/ 6, /*BUSY=*/
5));
//GxEPD2_BW<GxEPD2_583, MAX_HEIGHT(GxEPD2_583)>
display(GxEPD2_583(/*CS=*/ 4, /*DC=*/ 7, /*RST=*/ 6, /*BUSY=*/
5));
//GxEPD2_BW<GxEPD2_750, MAX_HEIGHT(GxEPD2_750)>
display(GxEPD2_750(/*CS=*/ 4, /*DC=*/ 7, /*RST=*/ 6, /*BUSY=*/
5));
```

The second half is more important to us because it includes the definition of the three-color e-ink modules, like the one we are using:

```
// 3-color e-papers
#define MAX_HEIGHT_3C(EPD) (EPD::HEIGHT <=
(MAX_DISPAY_BUFFER_SIZE / 2) / (EPD::WIDTH / 8) ? EPD::HEIGHT :
(MAX_DISPAY_BUFFER_SIZE / 2) / (EPD::WIDTH / 8))
//GxEPD2_3C<GxEPD2_154c, MAX_HEIGHT_3C(GxEPD2_154c)>
display(GxEPD2_154c(/*CS=*/ 4, /*DC=*/ 7, /*RST=*/ 6,
/*BUSY=*/ 5));
//GxEPD2_3C<GxEPD2_213c, MAX_HEIGHT_3C(GxEPD2_213c)>
display(GxEPD2_213c(/*CS=*/ 4, /*DC=*/ 7, /*RST=*/ 6,
/*BUSY=*/ 5));
//GxEPD2_3C<GxEPD2_290c, MAX_HEIGHT_3C(GxEPD2_290c)>
display(GxEPD2_290c(/*CS=*/ 4, /*DC=*/ 7, /*RST=*/ 6,
/*BUSY=*/ 5));
//GxEPD2_3C<GxEPD2_270c, MAX_HEIGHT_3C(GxEPD2_270c)>
display(GxEPD2_270c(/*CS=*/ 4, /*DC=*/ 7, /*RST=*/ 6,
/*BUSY=*/ 5));
// GxEPD2_3C<GxEPD2_420c, MAX_HEIGHT_3C(GxEPD2_420c)>
display(GxEPD2_420c(/*CS=*/ 4, /*DC=*/ 5, /*RST=*/ 3,
/*BUSY=*/ 2));
GxEPD2_3C<GxEPD2_420c, MAX_HEIGHT_3C(GxEPD2_420c)>
display(GxEPD2_420c(/*CS=*/ 4, /*DC=*/ 7, /*RST=*/ 6,
/*BUSY=*/ 5));
//GxEPD2_3C<GxEPD2_583c, MAX_HEIGHT_3C(GxEPD2_583c)>
```

```
display(GxEPD2_583c(/*CS=4*/ 4, /*DC=*/ 7, /*RST=*/ 6,
/*BUSY=*/ 5));
 //GxEPD2_3C<GxEPD2_750c, MAX_HEIGHT_3C(GxEPD2_750c)>
display(GxEPD2_750c(/*CS=4*/ 4, /*DC=*/ 7, /*RST=*/ 6,
/*BUSY=*/ 5));
 #endif
```

The library author prepared different versions for us for different e-paper modules. All lines are commented. We now need to find the one that is relevant to us and uncomment it.

The top part is relevant for black and white e-paper modules and the bottom one is for three-color e-paper modules. As we use a 4.2-inch three-color module, we need to uncomment the following line:

```
//GxEPD2_3C<GxEPD2_420c, MAX_HEIGHT_3C(GxEPD2_420c)>
display(GxEPD2_420c(/*CS=4*/ 4, /*DC=*/ 7, /*RST=*/ 6,
/*BUSY=*/ 5));
```

3. Delete the leading // so it looks like this:

```
GxEPD2_3C<GxEPD2_420c, MAX_HEIGHT_3C(GxEPD2_420c)>
display(GxEPD2_420c(/*CS=4*/ 4, /*DC=*/ 7, /*RST=*/ 6,
/*BUSY=*/ 5));
```

By looking at it closely, we can also see four more ports being mentioned:

- **CS**: 4
- **DC**: 7
- **RST**: 6
- **BUSY**: 5

These were the last pins missing for us to connect the e-paper module to the Arduino.

4. Connect all the pins like this (left is the e-paper panel pin, followed by the Arduino pin):
   - **BUSY**: 5
   - **RST**: 6
   - **DC**: 7
   - **CS**: 4
   - **CLK**: 9 (SCK)
   - **DIN**: 8 (MOSI)

- **GND**: GND
- **VCC**: VCC

Here is a diagram that shows how to connect the Arduino MKR WiFi 1010 to the Waveshare 4.2-inch three-color e-ink display module:

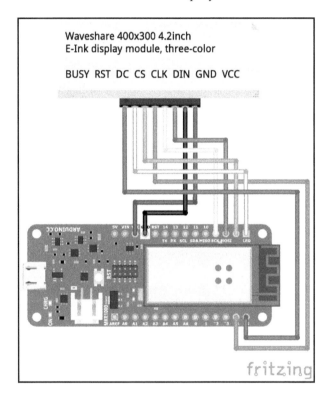

Waveshare e-ink display connected to an Arduino (this image was created with Fritzing)
License : (https://creativecommons.org/licenses/by-sa/3.0/)

Great. Let's see if this works.

5. Upload the code to your Arduino MKR WiFi 1010 (make sure the Arduino MKR WiFi 1010 is selected under **Tools** | **Board**, and the correct port under **Tools** | **Port**).

If the code is compiled correctly, the LED on the Arduino board should blink while the upload is in progress, and you should shortly see text displayed on the e-paper module:

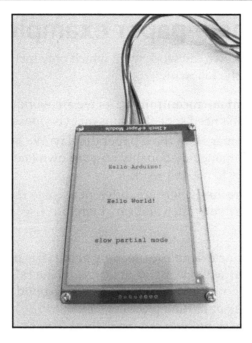

Example text shown on the e-ink display

Every time the content of the display is changed, you will see a few flashes. Don't worry, your e-paper module is not damaged. This is how e-paper modules change their content. Maybe you have used an eBook-reader based on e-paper before; they also need to change all pixels in order to display new content.

Now that we know the example code is working, we need to create our own sketch based on the example. Often, examples, as part of a library, are a good starting point. However, in this case, the example is extremely complex and uses a lot of functions that we don't need. So we need to clean it up a bit and delete all the code that is not relevant to us. We don't need code for different microcontrollers, we don't need to display images, we just need to display text, and therefore, we can remove 80% of the code, thereby simplifying our code base.

Instead of going through the example line by line and telling you what we need and what we can get rid of, I will tell you my strategy on how to simplify the code. Later on, I will present the simplified version of the code as a starting point for the project.

Simplifying the e-paper example

In order to create a code base we can work with, which only includes the code we really need, we have to get rid of the following:

- **Code for different microcontrollers**: As we are working with the Arduino MKR WiFi 1010, we don't need code that runs on every possible microcontroller.
- **Code to display images on the e-paper display**: We only want to display text. If you want to add image functionality on your own later on, you can re-integrate the bits and pieces to make it work.
- **Code for fast refreshes**: There are different ways to refresh the display. We don't update the display very often and can, therefore, use the slower but simpler method.

Deleting code from an existing example (or reading about it) is probably not the most exciting thing to do, but bear with me. Being able to combine existing examples and modify them to your needs is very powerful and will allow you to build complex projects without needing to code much. Think of putting pieces of a puzzle together—some pieces might need some pressure to connect, but you can make them fit with some extra work and create something new.

Now have a look at the Arduino IDE with the GxEPD2_Example example to see the changes I had to apply to simplify the code. You don't need to make the changes yourself because I won't go into detail with every change that I make. Instead, I will let you know when it is time to open my simplified code and jump back in.

Switch to the GxEPD2_boards_added.h tab. Here, you will find a lot of precompiler definitions that are necessary for different microcontrollers and display types. As the whole book is using the Arduino MKR WiFi 1010, I am pretty sure you also use it. Maybe the e-paper display is a little bit different, though. I am assuming you are using the same one I use, but if you don't, this code block is relevant to you:

```
#if defined(ARDUINO_ARCH_SAMD)
...
  //GxEPD2_BW<GxEPD2_154, MAX_HEIGHT(GxEPD2_154)> display(GxEPD2_154(...));
  //GxEPD2_BW<GxEPD2_213, MAX_HEIGHT(GxEPD2_213)> display(GxEPD2_213(...));
  //GxEPD2_BW<GxEPD2_213_flex, MAX_HEIGHT(GxEPD2_213_flex)>
display(GxEPD2_213_flex(...));
  //GxEPD2_BW<GxEPD2_290, MAX_HEIGHT(GxEPD2_290)> display(GxEPD2_290(...));
  //GxEPD2_BW<GxEPD2_290_T5, MAX_HEIGHT(GxEPD2_290_T5)>
display(GxEPD2_290_T5(...));
  //GxEPD2_BW<GxEPD2_270, MAX_HEIGHT(GxEPD2_270)> display(GxEPD2_270(...));
  //GxEPD2_BW<GxEPD2_420, MAX_HEIGHT(GxEPD2_420)> display(GxEPD2_420(...));
  //GxEPD2_BW<GxEPD2_583, MAX_HEIGHT(GxEPD2_583)> display(GxEPD2_583(...));
```

```
//GxEPD2_BW<GxEPD2_750, MAX_HEIGHT(GxEPD2_750)> display(GxEPD2_750(...));
// 3-color e-papers
...
//GxEPD2_3C<GxEPD2_154c, MAX_HEIGHT_3C(GxEPD2_154c)>
display(GxEPD2_154c(...));
//GxEPD2_3C<GxEPD2_213c, MAX_HEIGHT_3C(GxEPD2_213c)>
display(GxEPD2_213c(...));
//GxEPD2_3C<GxEPD2_290c, MAX_HEIGHT_3C(GxEPD2_290c)>
display(GxEPD2_290c(...));
//GxEPD2_3C<GxEPD2_270c, MAX_HEIGHT_3C(GxEPD2_270c)>
display(GxEPD2_270c(...));
GxEPD2_3C<GxEPD2_420c, MAX_HEIGHT_3C(GxEPD2_420c)>
display(GxEPD2_420c(...));
//GxEPD2_3C<GxEPD2_583c, MAX_HEIGHT_3C(GxEPD2_583c)>
display(GxEPD2_583c(...));
//GxEPD2_3C<GxEPD2_750c, MAX_HEIGHT_3C(GxEPD2_750c)>
display(GxEPD2_750c(...));
#endif
```

We are using the 4.2-inch e-paper module with three colors, so for us, `GxEPD2_420c` is the one to use. All the others are commented, and therefore, unused. We can get rid of all commented lines, as well as the precompiler definitions targeted at other microcontrollers. This leaves us with the following lines:

```
#if defined(ARDUINO_ARCH_SAMD)
#define MAX_DISPAY_BUFFER_SIZE 15000ul // ~15k is a good compromise
#define MAX_HEIGHT(EPD) (EPD::HEIGHT <= MAX_DISPAY_BUFFER_SIZE /
(EPD::WIDTH / 8) ? EPD::HEIGHT : MAX_DISPAY_BUFFER_SIZE / (EPD::WIDTH / 8))
#define MAX_HEIGHT_3C(EPD) (EPD::HEIGHT <= (MAX_DISPAY_BUFFER_SIZE / 2) /
(EPD::WIDTH / 8) ? EPD::HEIGHT : (MAX_DISPAY_BUFFER_SIZE / 2) / (EPD::WIDTH
/ 8))
GxEPD2_3C<GxEPD2_420c, MAX_HEIGHT_3C(GxEPD2_420c)>
display(GxEPD2_420c(/*CS=4*/ 4, /*DC=*/ 7, /*RST=*/ 6, /*BUSY=*/ 5));
#endif
```

Because we already know that we are working with the Arduino MKR WiFi 1010 (and not another microcontroller), we can also get rid of the `#if`/`#endif` precompiler statements, leaving us with just a few lines of code:

```
#define MAX_DISPAY_BUFFER_SIZE 15000ul // ~15k is a good compromise
#define MAX_HEIGHT(EPD) (EPD::HEIGHT <= MAX_DISPAY_BUFFER_SIZE /
(EPD::WIDTH / 8) ? EPD::HEIGHT : MAX_DISPAY_BUFFER_SIZE / (EPD::WIDTH / 8))
#define MAX_HEIGHT_3C(EPD) (EPD::HEIGHT <= (MAX_DISPAY_BUFFER_SIZE / 2) /
(EPD::WIDTH / 8) ? EPD::HEIGHT : (MAX_DISPAY_BUFFER_SIZE / 2) / (EPD::WIDTH
/ 8))
GxEPD2_3C<GxEPD2_420c, MAX_HEIGHT_3C(GxEPD2_420c)>
display(GxEPD2_420c(/*CS=4*/ 4, /*DC=*/ 7, /*RST=*/ 6, /*BUSY=*/ 5));
```

This already looks much better, right?

We can move this code into our main sketch file, leaving us with an empty GxEPD2_boards_added.h tab, which is no longer needed and can, therefore, be deleted.

When merging different examples/sketches, or moving code from one position to another, it is important to consider a few things:

1. The #define statements and global variable definitions should be placed at the top of the sketch, taking into account that any libraries being used are included before them. In our case, the code must be placed after the library was included, so after the #include <GxEPD2_3C.h> line but before any other code that uses display.

2. There can only be one loop() and one setup() function, so you need to merge them individually (as we have done before). Also, you need to make sure that all the variables used inside these functions are defined before (at the top of the sketch).

3. When merging different sketches and one of them contains delay() calls, halting the program for a certain amount of time, the other parts of the program might stop working. This is because delay() basically tells the microcontroller not to do anything, so any code that depends on being executed periodically, for example, might be disturbed by this. MQTT is a good example of this. If the client does not ping the MQTT server in a certain amount of time, it will be disconnected. So, often it is better to rewrite delay() calls to use a custom timer instead, using the millis() function.

After deleting the GxEPD2_boards_added.h file, we also have to get rid of the import statement in our main sketch file; otherwise, we would get a compile error because the file cannot be found anymore. This line must be deleted:

```
#include "GxEPD2_boards_added.h"
```

Let's move on to clean up the main file (GxEPD2_Example).

Have a look at the setup function:

```
void setup()
{
    Serial.begin(115200);
    Serial.println();
    Serial.println("setup");
    delay(100);
```

```
display.init(115200);
// first update should be full refresh
helloWorld();
delay(1000);
// partial refresh mode can be used to full screen,
// effective if display panel hasFastPartialUpdate
helloFullScreenPartialMode();
delay(1000);
helloArduino();
delay(1000);
helloEpaper();
delay(1000);
```

In the second part of the setup function, even more demo functions are called:

```
showFont("FreeMonoBold9pt7b", &FreeMonoBold9pt7b);
delay(1000);
drawBitmaps();
if (display.epd2.hasPartialUpdate)
{
    showPartialUpdate();
    delay(1000);
} // else // on GDEW0154Z04 only full update available, doesn't look
nice
//drawCornerTest();
//showBox(16, 16, 48, 32, false);
//showBox(16, 56, 48, 32, true);
display.powerOff();
deepSleepTest();
Serial.println("setup done");
}
```

The following is a list of all the custom functions that are called from our setup function:

- helloWorld()
- helloFullScreenPartialMode()
- helloArduino()
- helloEpaper()
- drawBitmaps()
- drawCornerTest()
- showBox(...)
- deepSleepTest()

When you restart the Arduino (by unplugging the USB cable and plugging it back in, or pressing the reset button twice on the Arduino), the first screen you will see is the hello world screen. It displays the `Hello World` string. This is all we need. So we can delete all other functions (and their calls inside the `setup` function), leaving us with just three defined functions: `setup`, `loop`, and `helloWorld`.

There are also some precompiler statements that we need. The first part contains all precompiler statements (for example, `#define` and `#include`) as well as global variable definitions (for example, `const char HelloWorld[] = "Hello World";`). After deleting everything else and changing the code style a bit due to personal preference, we end up with the following code:

```
#include <GxEPD2_BW.h>
#include <GxEPD2_3C.h>
#include <Fonts/FreeMonoBold9pt7b.h>

#define MAX_DISPLAY_BUFFER_SIZE 15000ul
#define MAX_HEIGHT(EPD) (EPD::HEIGHT <= MAX_DISPLAY_BUFFER_SIZE /
(EPD::WIDTH / 8) ? EPD::HEIGHT : MAX_DISPLAY_BUFFER_SIZE / (EPD::WIDTH / 8))
#define MAX_HEIGHT_3C(EPD) (EPD::HEIGHT <= (MAX_DISPLAY_BUFFER_SIZE / 2) /
(EPD::WIDTH / 8) ? EPD::HEIGHT : (MAX_DISPLAY_BUFFER_SIZE / 2) / (EPD::WIDTH
/ 8))
 GxEPD2_3C<GxEPD2_420c, MAX_HEIGHT_3C(GxEPD2_420c)>
display(GxEPD2_420c(/*CS=4*/ 4, /*DC=*/ 7, /*RST=*/ 6, /*BUSY=*/ 5));

const char HelloWorld[] = "Hello World";
```

Directly after this, there is the `setup` function:

```
void setup() {
    Serial.begin(115200);
    Serial.println("starting...");
    display.init(115200);
    helloWorld();
    display.powerOff();
}
```

It first initializes the serial port with the baud rate (`115200`) so we can send and receive messages via the Arduino serial monitor (in this case, `starting...`). We then initialize the display unit with the same baud rate, run the `helloWorld` function, which we will have a look at in a bit, and afterwards, we turn the display off. Turning the display off ensures that it consumes less power. Because it is not a regular display (which would turn black) but an e-paper display, text and images displayed on the screen remain visible even without power. This is what makes e-paper displays (which use e-ink technology) so special.

After the `setup` function, we define the `loop` function:

```
void loop() {}
```

It is empty and doing nothing for now. We just display the `Hello World` text on the screen, initiated from the `setup` function, and that's it.

After the `loop` function, we have the `helloWorld` function, which we copied over unchanged from `GxEPD2_Example`. As we have seen in the `setup` function, it is being called directly after the display was initiated. We could have put all the code inside the `setup` function as well, but this way, we have a better separation. The `helloWorld` function looks like this:

```
void helloWorld() {
    display.setRotation(1);
    display.setFont(&FreeMonoBold9pt7b);
    display.setTextColor(GxEPD_BLACK);
    int16_t tbx, tby; uint16_t tbw, tbh;
    display.getTextBounds(HelloWorld, 0, 0, &tbx, &tby, &tbw, &tbh);
    uint16_t x = (display.width() - tbw) / 2;
    uint16_t y = (display.height() + tbh) / 2; // y is base line!
    display.setFullWindow();
    display.firstPage();
    do {
        display.fillScreen(GxEPD_WHITE);
        display.setCursor(x, y);
        display.print(HelloWorld);
    } while (display.nextPage());
}
```

Let's look at it line by line:

```
display.setRotation(1);
```

This puts the display in portrait mode:

```
display.setFont(&FreeMonoBold9pt7b);
```

The display can show text with different fonts. You can use serif, sans-serif, or monospace fonts. If you have a closer look at the font name, you can see 9pt, which implies that font file displays the text in 9 point (**pt** is the abbreviation for **point**).

In common text editors on macOS, Linux, and Windows, font files are vector based and you can use the same font file to display text in 9 points or 90 points. Vectors scale infinitely.

On microcontrollers, we don't have this luxury. The Arduino MKR WiFi 1010, and most other microcontrollers have very limited memory, and therefore, we are not able to use the same font files as we would on our desktop computers.

The `GxEPD2` library uses `Adafruit_GFX`, a library by the manufacturer and webshop Adafruit (`https://adafruit.com`). It is used to display images on the screen and render text. Unlike the desktop fonts, we have to load special versions of fonts that are based on pixels. So there is one font that can be used to display text in 9 points, and another to display text in 90 points.

Let's have a look at the rest of the font names we currently use (`FreeMonoBold9pt7b`): `FreeMono` is the name of the font, followed by the style **Bold**, **Oblique** (italic), or none for the regular version. The third part of the name is the size of the font, in this case, 9 points, followed by `7b`, which stands for 7 bit, and defines how many bits are used to store one character in the font. You can ignore this part of the font name.

To get a complete overview of the fonts in the Adafruit GFX library, you can have a look at the `Adafruit_GFX` repository on GitHub (`https://github.com/adafruit/Adafruit-GFX-Library/tree/master/Fonts`). Please note that the version you have installed via the Arduino Library Manager might be different to the one shown on GitHub and it, therefore, could be that not all the fonts are available on your computer. If you want to know for sure, you can open your Arduino library folder on your hard drive and navigate to the `Adafruit_GFX` fonts folder. On macOS, this folder can be found at `/Users/YOUR_NAME/Documents/Arduino/libraries/Adafruit_GFX_Library/Fonts`, and on Windows you can find it at `My Documents\Arduino\libraries\Adafruit_GFX_Library\Fonts`.

If you want to learn more about how fonts are displayed on the e-paper display and other displays on the Arduino (such as LED and OLED), you can read the detailed Adafruit guide (`https://learn.adafruit.com/adafruit-gfx-graphics-library/using-fonts`).

Okay, let's get back to our `helloWorld` function. There are a few more lines that we need to discuss:

```
display.setTextColor(GxEPD_BLACK);
int16_t tbx, tby; uint16_t tbw, tbh;
display.getTextBounds(HelloWorld, 0, 0, &tbx, &tby, &tbw, &tbh);
uint16_t x = (display.width() - tbw) / 2;
uint16_t y = (display.height() + tbh) / 2; // y is base line!
display.setFullWindow();
display.firstPage();
do {
    display.fillScreen(GxEPD_WHITE);
    display.setCursor(x, y);
```

```
        display.print(HelloWorld);
    } while (display.nextPage());
```

`display.setTextColor(GxEPD_BLACK);` sets the color of the text (surprise, surprise). Depending on the kind of display, we can use different values here. The display we are using supports three colors: black, white, and red. All of them are defined by the GxEPD2 library and can be specified by using the GxEPD_BLACK, GxEPD_WHITE, and GxEPD_RED variables.

The next four lines look more cryptic and are used to center the text on the display:

```
int16_t tbx, tby; uint16_t tbw, tbh;
display.getTextBounds(HelloWorld, 0, 0, &tbx, &tby, &tbw, &tbh);
uint16_t x = (display.width() - tbw) / 2;
uint16_t y = (display.height() + tbh) / 2;
```

We first define variables to store the text boundaries (imagine a rectangle around the text we want to display). This imaginary rectangle has four properties: the position (x/y, stored in tbx/tby), width (tbw), and height (tbh).

We then call the getTextBounds function of the display object, passing it the text we want to display ("Hello World", which is stored in the HelloWorld variable), 0 twice which is only relevant for other uses of the function, followed by references to the variables we just created to store the x and y position of the text boundary, and finally, the width and height of the text boundary.

After this, tbw contains the width of the text that is about to be displayed in pixels. tbh then contains the height of the text. We don't use the position tbx or tby, but we needed to pass it to the display.getTextBounds function anyway, otherwise, it would throw a compile error.

If you feel like this is overly complicated, I totally agree with you. Having a display.getTextWidth(HelloWorld) and display.getTextHeight(HelloWorld) function would be nice. In Germany, you say "*Das Leben ist kein Ponyhof*" ("*Life is not a pony farm*"), meaning that life is not always easy/not everything goes the way you want it to go. We just have to deal with it.

The next line (`display.setFullWindow()`) defines that we want to use the whole available area of the display. The last parts looks a bit cryptic again:

```
display.firstPage();
do {
    display.fillScreen(GxEPD_WHITE);
    display.setCursor(x, y);
    display.print(HelloWorld);
} while (display.nextPage());
```

The `display.firstPage()` and `while (display.nextPage())` lines are needed to pass on the content from the Arduino MKR WiFi 1010 to the buffer of the display. There are just two things that you need to know here:

- Everything you want to render (text/images) needs to be inside the curly braces (between `do {` and the closing `}  while  (...)`).
- You should not read in any analog sensor values in this code block. According to the library author, it might lead to strange side effects.

The three lines of code inside the curly braces are used for the actual drawing. First, we set the background color of the display to white by calling `display.fillScreen(GxEPD WHITE)`. We then set the cursor position to the position we calculated earlier. Last but not least we call `display.print(HelloWorld)`. The `HelloWorld` parameter is a variable that we defined at the beginning of our sketch (`const char HelloWorld[] = "Hello World"`).

In order to upload the code now, you have to first open the simplified example code:

1. If you haven't done so already, please go to `https://github.com/PacktPublishing/Hands-on-Internet-of-Things-with-MQTT` and download the source code for the book by clicking **Clone or download**, then **Download Zip**.
2. Once the ZIP has finished downloading, please extract it, navigate to `ch6/arduino`, and open the `ch6_01_example_reduced` Arduino sketch.
3. Save it as a new version, so if there are any errors, you can come back to the original version and compare it to yours. Save your copy of the sketch as `ch6_my_code`.

From now on, please follow along by making the changes to your code as well. I will save versions on the way (checkpoints), so you always have code to go back to if you run into problems, which will ensure you don't have to start from the beginning if there were any issues.

Now upload the code to your Arduino and you should see "Hello World" on your display. If you see any errors, several things might have gone wrong:

- The GxEDP2 or Adafruit_GFX libraries were not installed correctly.
- The Arduino MKR WiFi 1010 was not selected under **Tools | Board**.
- The right serial port was not selected under **Tools | Port**.
- The USB cable is not connected.

If you could upload the code but you don't see Hello World on the screen, one of the following things could have gone wrong:

- The cables are connected in the wrong way—please check the correct mapping at the beginning of the chapter. This is the most likely cause of the error.
- One (or multiple) of the cables are loose/don't have a proper connection.
- One of the cables, or jumper-cables if you used them to extend the cables, is broken. It is quite unlikely that the cables coming from the display are broken. But jumper-cables sometimes really are broken. In this case, try replacing the jumper cables.

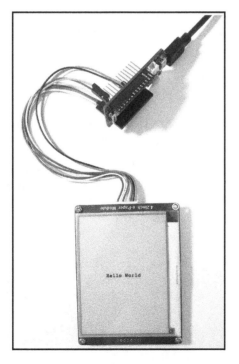

Hello World text showing on the e-ink display

I assume everything worked out fine and you can see `Hello World` on the screen of your display. Great!

Modifying the e-paper example

Now we need to do a couple of changes to the code. Currently the `helloWorld` function always prints the same text to the display. We want to make it dynamic, first by sending text via the serial port (using the Arduino serial monitor), then via MQTT. We will also make the text bigger by using another font file, rotate the text (so we can display longer words/phrases) and add the possibility to color the text red.

Our e-paper project can be used in various ways, but the main purpose is to attach it to your wall near the flat/house entry door and send text to it via your smartphone or desktop computer. This way, you can remind yourself to bring out the trash or buy groceries and you see it whenever you walk past the door. If the thing you want to remind yourself of is not super important, the text will be black. If it really is important, it will be red. To figure out if something is important or not, we could use machine learning, but that might be a bit overkill here. So let's do it manually by implementing a check: if the text we send to the Arduino MKR WiFi 1010 ends with an exclamation mark (`!`), it must be important and therefore colored red. If the last character is not an exclamation mark, it should be displayed in black.

Once we have that done, we will use either an Android or iOS app to send messages to our device. This way, we can update the text on the display even when we are not at home.

If you do not want to use the device as a task-reminder for yourself, you can also give it to your partner and send them nice messages when you are not with them. Your choice.

To get all that working, we still have some work to do. Let's start by re-writing the `helloWorld` function so that we can display arbitrary text and not always show `Hello World`.

Because we want it to be dynamic, `helloWorld` is not a very good name for this:

1. Change `void helloWorld() {` to `void setText() {`.

   Because we call this function in our `setup` function, we also need to change it there.

2. Inside the `setup` function, change `helloWorld();` to `setText();`.

   Now we want to add a parameter to the function so we can pass in different text to the function.

3. Add a string parameter to the function by changing `void setText() {` to `void setText(String text) {`. Here, we will pass in the text to be displayed.

   We also want to make use of it and replace all places where we use the `HelloWorld` variable right now.

4. Delete the `HelloWorld` variable (as we don't need it anymore) by removing the following line:

   ```
   const char HelloWorld[] = "Hello World";
   ```

5. Change `display.getTextBounds(...)` to the following:

   ```
   display.getTextBounds(text, 0, 0, &tbx, &tby, &tbw, &tbh);
   ```

   So it uses the parameter of our function instead of the `HelloWorld` variable we just deleted. We also need to change the call where we actually define which text to print:

   ```
   display.print(HelloWorld);
   ```

6. Change the preceding line to this:

   ```
   display.print(text);
   ```

7. Compile the code now by pressing the button next to the upload button (the left-most button). You will get a compiler error:

   ```
   too few arguments to function 'void setText(String)'
   ```

   In the `setup` function, we call `setText();`, but our `setText` function expects a parameter of type string containing the actual text. Let's change that by adding some text to our function call.

8. Change `setText();` inside the `setup` function to this:

   ```
   setText("Hello display");
   ```

9. Compile again, upload the code, and the text on the display should change to `Hello display`.

Now let's make better use of the display by changing it from portrait mode to landscape mode. This is an easy one. Do you have any idea how to do it?

The first line in our `setText` function defines the rotation. Currently it is this:

```
display.setRotation(1);
```

10. Change it to the following:

```
display.setRotation(0);
```

11. Re-upload the code, and the text will be shown in landscape mode.

Let's move on to send a dynamic text to the display.

Making your e-paper device accessible via serial

The next thing we want to do is make the text on the screen dynamic by adding serial communication. At the beginning, the text was static and defined as the `HelloWorld` variable. Now we have changed it to be a dynamic parameter of our `setText` function, but the way we use it is still static, because we just pass the `"Hello display"` static text to it via `setText("Hello display");`.

To make use of the serial port, let's find an example first that reads in a string from the serial port. We will then integrate it into our sketch. All the serial port examples can be found under **File** | **Examples** | **04.Communication**.

Sadly, there is none that receives an input string and then does something with it.

We can have a look at the Arduino serial reference (`https://www.arduino.cc/reference/en/language/functions/communication/serial/`) to see if there is a function that does what we need. And indeed there is—`readStringUntil()`. According to the reference, it reads characters from the serial buffer into a string. Exactly what we need, great!

Serial communication is character by character. The concept of lines does not exist. If inside the serial monitor we enter some random text, let's say `hello`, and press *Enter* to send it, under the hood, additional characters will be sent that you cannot see. We will make use of the hidden \n character (which represents a new line). It is added to the string once we press **Send**.

To try this out, we need to create a new sketch. The goal of this mini-sketch is to be able to send strings via serial and echo them. Once this is working, we will include the code in our main sketch. Refer to the following steps to get started:

1. Create a new sketch, name it `serial_string_echo`, and paste the following code:

```
void setup() {
    Serial.begin(115200);
}

void loop() {
    if(Serial.available() > 0) {
        String text = Serial.readStringUntil('\n');
        Serial.print("Received: "); Serial.println(text);
    }
}
```

2. Upload the sketch and open the Arduino serial monitor (**Tools** | **Serial Monitor**).
3. Enter some text and either press the **Send** button or the *Enter* key. It will be sent to the Arduino via the serial interface, as shown in the following screenshot:

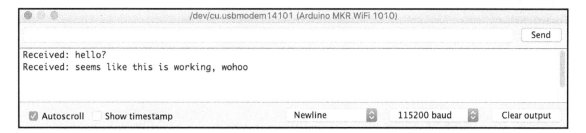

The Arduino serial monitor

Please make sure that **Newline** is selected in the first select field, and `115200` baud in the second one.

In our code, we will wait as long as it takes until we find an \n character inside the serial stream because of this line: `Serial.readStringUntil('\n');`. So, make sure **Newline** is selected in the serial monitor. Otherwise, your Arduino will wait until the function times out.

Once a string is found that included the newline character, it will be echoed as follows:

```
Serial.print("Received: "); Serial.println(text);
```

 The `println` function also make use of this hidden character. `println` behaves the same way as `print`, but adds the hidden newline character to the text.

The other thing that is important to set up correctly in the Arduino serial monitor is the baud rate (the number on the bottom right of the window). This must match the rate in our `Serial.begin` call in our sketch.

In my case, I am using `Serial.begin(115200);` in my sketch, and my Arduino serial monitor uses `115200` as well. Entering messages in the text field and sending them results in a `Received: xyz` echo from the Arduino.

If you run into any problems here, try restarting the Arduino IDE or unplugging the Arduino and plugging it back in again (you might need to reset the serial port).

Let's integrate this serial code into our main sketch so we can send commands via the serial interface to the Arduino and therefore change the text on the display. Follow these instructions:

1. To initialize the serial interface, we need to include the following line in our `setup` function:

   ```
   Serial.begin(115200);
   ```

2. By looking at our existing `setup` function, we can see that this is already here:

   ```
   void setup() {
       Serial.begin(115200);
       Serial.println("starting...");
       display.init(115200);
       setText("Hello display");
       display.powerOff();
   }
   ```

3. Let's move on to the `loop` function. Currently, in our main sketch, it is empty because we only show the `Hello display` text and nothing more. Our current `loop` function looks like this:

   ```
   void loop() {}
   ```

4. Let's change this by replacing it with the `loop` function of our serial test sketch. After replacing the empty one in our main sketch, it should look like this:

```
void loop() {
    if(Serial.available() > 0) {
        String text = Serial.readStringUntil('\n');
        Serial.print("Received: "); Serial.println(text);
    }
}
```

Let me explain what is happening here. Every time the `loop` function is executed, we check if there are any characters waiting for us on the serial interface. `Serial.available()` returns the number of characters available, so by checking `Serial.available() > 0` inside the `if` clause, we make sure the code inside the `if` clause is only executed if there is at least one character available.

The following line waits until it finds a newline character (\n) in the serial stream. This is a blocking function. It will wait as long as it takes for the \n character to arrive. `delay()` is also a blocking function. Using blocking functions such as `delay()` or `readStringUntil()` can lead to problems in your code because they put the Arduino into an idle state until a condition is fulfilled. In the case of `delay`, it is *be idle until x milliseconds are over*. In the case of `readStringUntil`, it is *don't do anything until the newline character arrives*.

So, blocking functions should be used with caution. In our case, we just want to use it for testing purposes, so it is okay to use it. Later on, we will implement communication via MQTT anyway and not use the serial interface anymore.

If you want to learn more about best practices using the serial interface, you should read the excellent write-up by Arduino user Robin2 (http://forum.arduino.cc/index.php?topic=396450.0). It will explain blocking versus non-blocking serial usage and provide examples for common use-cases.

Now that you know `readStringUntil` and that blocking functions can lead to problems, you might ask yourself why I still recommend using it in our code. The answer is simplicity. Doing it the proper way using non-blocking code would require many more lines of code, which would over-complicate it for our use case.

The `Serial` function is not of much use so far. It waits until it receives a newline character and then sends it back to us. To display the text we send via the Arduino serial monitor, all we need to do is add one more line of code to pass on the text we just read into our `setText` function.

5. Add the `setText` function call after the `Serial.print` statement so our `loop` function looks like this:

```
void loop() {
    if(Serial.available() > 0) {
        String text = Serial.readStringUntil('\n');
        Serial.print("Received: "); Serial.println(text);
        setText(text);
    }
}
```

Before we upload the code, let's make one more change. Because we are using the display in landscape mode, we have plenty of space left. Let's use a bigger font so it is easier to read the text.

Earlier, we had a look at how font loading works, especially which fonts are available when using the `Adafruit_GFX` library.

Currently we use the `FreeMonoBold9pt7b` font, a 9-pixel monospace font. Let's use a sans-serif one instead with a size of 18 instead of 9 pixels: `FreeSansBold18pt7b`.

To change the font being used, we need to change two lines of code: first, the `#include` statement at the beginning of the sketch, and second, the line where we tell the library which font to use.

6. Replace the following line:

```
#include <Fonts/FreeMonoBold9pt7b.h>
```

This line should be used as a replacement:

```
#include <Fonts/FreeSansBold18pt7b.h>
```

7. Now scroll down to our `setText` function to the following line:

```
display.setFont(&FreeMonoBold9pt7b);
```

The line as follows should become its replacement:

```
display.setFont(&FreeSansBold18pt7b);
```

Feel free to try any of the other available fonts later on. You just need to replace these two lines of code.

8. Upload the code.

The first thing that you will notice is the bigger font. It looks better, doesn't it?

Now let's try if we can send text from the computer to the Arduino MKR WiFi 1010, which will then pass it on to the e-paper module and display it.

9. Open the Arduino serial monitor (**Tools** | **Serial Monitor**), enter some text in the input box, and press **Send** (or the *Enter* key).

Tadaa! Your e-paper display should now refresh and display the text you sent to it.

Now let's move on and add an option to display very important text. Often, important phrases/words end with an exclamation mark (!). So we could simply create a rule like this: if text ends with an exclamation mark, set the text color to red; otherwise, use black.

Right now, we always set the text color to black using this line inside our `setText` function:

```
display.setTextColor(GxEPD_BLACK);
```

10. Replace the previous line with the following lines:

```
char lastCharacter = text.charAt(text.length() - 1);
if (lastCharacter == '!') {
    display.setTextColor(GxEPD_RED);
} else {
    display.setTextColor(GxEPD_BLACK);
}
```

The first line extracts the last character from the text. For this, we use the `charAt` function of the string class.

You should have a look at the Arduino string class reference (https://www.arduino.cc/reference/en/language/variables/data-types/string/) to find out what other functions it offers.

When working with Arduino, some people prefer using character arrays instead of the string class because it is more efficient. It definitely is, but it also adds complexity to your code. If simplicity is more important to you than performance and efficiency, you should use the string class in your projects as we do here as well.

The parameter we pass into the `charAt` function is the last index of the text. If we, for example, send the `Hello` text to our device, the character indices will look like this: `H: 0, e: 1, l: 2, l: 3, o: 4`. The string contains five characters, so calling the `length` function will return `5`. Subtracting one from it gives us the last character in the string.

In the `if/else` block, we check whether this character is an exclamation mark. If it is, we set the text color to red, otherwise, we set it to black.

11. Re-upload your code.
12. Open the Arduino serial monitor, enter some text with an exclamation mark at the end, and send it over to your Arduino.

That was easy, wasn't it? You should now see the text that you just entered appears on the e-paper display, as shown in the following photo:

Display showing the very important text in red

We are all set now to finally include MQTT communication so we can send our very important to-do items from our mobile phone via a third-party MQTT app.

You can find the code until here as checkpoint under the name `ch6_02_serial` in the downloaded source code repository in the `ch6 | arduino` folder.

If you have had any problems so far and could not make your code work, feel free to download this version instead. If you are also having problems with the downloaded version, there must be something wrong with either the installed libraries, the connection between Arduino and the e-paper module, or you have bad luck and one of them is damaged (unlikely). In any case, don't hesitate to open an issue on the book's repository on GitHub and we will figure it out together.

Preparing the MQTT integration

Before we integrate the MQTT code into our sketch, we first need to check our MQTT connection by following these steps:

1. Please open the `mqtt_shiftr_send_receive_example` example, which you will find in the `general | arduino` folder inside the downloaded source code repository.

    The example, as well as all other MQTT code running on the Arduino MKR WiFi 1010, depends on two libraries:

    - **WiFiNINA**: This provides network functionality to the Arduino MKR WiFi 1010.
    - **MQTT (by Joel Gaehwiler)**: This enables your Arduino to communicate with other devices and servers via MQTT.

    You should have these installed already. If not, install them via the Arduino Library Manager. I am using WiFiNINA v1.30 and MQTT v2.4.3.

    Before you can upload the code, you need to set your Wi-Fi name and password, which can often be found on the bottom side of your Wi-Fi router.

2. Replace `name` and `password` with your actual Wi-Fi username and password:

    ```
    const char WIFI_SSID[] = "name";
    const char WIFI_PASSWORD[] = "password";
    ```

3. Upload the code.
4. Open the Arduino serial monitor, and after a few seconds, you should see the following output:

    ```
    incoming: /tims-channel - hello
    incoming: /tims-channel - hello
    incoming: /tims-channel - hello
    incoming: /tims-channel - hello
    ```

We already had a closer look at the source code of this example in Chapter 5, *Building Your Own Automatic Pet Food Dispenser*, so we will not go into great detail about what exactly is happening in the code.

In a nutshell, we connect to your local Wi-Fi network, then connect to the MQTT server shiftr.io. Afterwards, we subscribe to the `/tims-channel` topic.

We also send out an MQTT message every second to the same channel containing the `hello` payload.

Because we are subscribed to the same channel, we also receive it, and `incoming: /tims-channel - hello` is printed in the serial monitor.

The `/tims-channel` channel name is freely picked, as well as the MQTT device name (in my case, `hellomqtt`):

```
const char MQTT_DEVICE_NAME[] = "hellomqtt";
```

If you see the previous serial output, you can be certain that your Wi-Fi connection, as well as the connection to the MQTT server shiftr.io, is working and that you are ready to move on to the next step—making your e-paper to-do device ready to be used with MQTT, so you can control it from wherever you are via smartphone or computer.

Making your e-paper device accessible with MQTT

Integrating MQTT into our existing code is easy. We already have all the pieces ready:

- Code to display a text on the e-paper module (our main sketch)
- An example to send text via MQTT

We now need to combine these two so the text we receive via MQTT on the Arduino is used to set the text on the display. Because we have a good abstraction of our code with the `setText` function, we only have to add one more line of code after integrating the code from the example.

The MQTT example consists of the following parts:

- Pre-compiler definitions and variable declarations (at the top of the sketch).
- `void setup()`: Called once at the beginning, this initializes the serial port and calls the Wi-Fi/MQTT connect function.
- `void loop()`: The function that is executed repeatedly after the `setup` function has finished.
- `void connect()`: Establishes a connection to the Wi-Fi router and then with the MQTT server.
- `void messageReceived`: Called when there is a new incoming MQTT message (to a topic we subscribed to).

Please follow along with these steps to integrate the MQTT example into our main sketch:

1. Copy the `connect` and `messageReceived` functions from the MQTT example and paste them at the end of your main sketch. There are no existing functions with this name, and therefore, it does not clash.

   First, copy and paste the `connect` function:

   ```
   void connect() {
     // first connect to the wifi
     Serial.print("Checking wifi...");
     while (status != WL_CONNECTED) {
       status = WiFi.begin(WIFI_SSID, WIFI_PASSWORD);
       Serial.print(".");
       delay(1000);
     }
     Serial.println(); Serial.print("Connected to WiFi!");
   Serial.println(); // second connect to the MQTT server
     client.begin(MQTT_SERVER, MQTT_SERVER_PORT, net);
     Serial.println("Connecting to MQTT server...");
     while (!client.connect(MQTT_DEVICE_NAME, MQTT_USERNAME,
   MQTT_PASSWORD)) {
       Serial.print(".");
       delay(1000);
     }
     Serial.println("Connected to MQTT server");
     // define what should happen when messages are incoming
     client.onMessage(messageReceived);
     // subscribe to MQTT topics
     client.subscribe("/tims-channel");
   }
   ```

   Then directly after, paste the `messageReceived` function as follows:

   ```
   void messageReceived(String &topic, String &payload) {
     Serial.println("incoming: " + topic + " - " + payload);
   }
   ```

2. Now also copy the precompiler statements (starting with #) and variable definitions from the MQTT example and paste them at the beginning of our main sketch.

> In general, when copying variable definitions and precompiler statements from one sketch to another, you should make sure that there are no duplicate lines. This can happen, for example, when both sketches import the same library. In our case, there are no duplicate lines.

3. Group the `#include` statements at the very top.

After copying over the precompiler statements and variable definitions, the first part of your code should look like the following. First we have the `#include` and `#define` statements:

```
#include <GxEPD2_BW.h>
#include <GxEPD2_3C.h>
#include <Fonts/FreeSansBold18pt7b.h>
#include <WiFiNINA.h>
#include <MQTT.h>

#define MAX_DISPAY_BUFFER_SIZE 15000ul
#define MAX_HEIGHT(EPD) (EPD::HEIGHT <= MAX_DISPAY_BUFFER_SIZE
/ (EPD::WIDTH / 8) ? EPD::HEIGHT : MAX_DISPAY_BUFFER_SIZE /
(EPD::WIDTH / 8))
#define MAX_HEIGHT_3C(EPD) (EPD::HEIGHT <=
(MAX_DISPAY_BUFFER_SIZE / 2) / (EPD::WIDTH / 8) ? EPD::HEIGHT :
(MAX_DISPAY_BUFFER_SIZE / 2) / (EPD::WIDTH / 8))
GxEPD2_3C<GxEPD2_420c, MAX_HEIGHT_3C(GxEPD2_420c)>
display(GxEPD2_420c(/*CS=4*/ 4, /*DC=*/ 7, /*RST=*/ 6,
/*BUSY=*/ 5));
```

These statements are followed by variable definitions, as shown here:

```
const char WIFI_SSID[] = "name"; // set your network name here
const char WIFI_PASSWORD[] = "password"; // set your network
password here
const char MQTT_SERVER[] = "broker.shiftr.io";
const int MQTT_SERVER_PORT = 1883;
const char MQTT_USERNAME[] = "try";
const char MQTT_PASSWORD[] = "try";
const char MQTT_DEVICE_NAME[] = "hellomqtt"; // can be freely
picked

int status = WL_IDLE_STATUS;
WiFiClient net;
MQTTClient client;

unsigned long lastMillis = 0;
```

Now we need to integrate the `loop` and `setup` functions. As these already exist, we need to integrate the code inside them and cannot simply copy them over completely.

Let's start with the `loop` function. The `loop` function of the MQTT example looks like this:

```
void loop() {
  client.loop();
  if (!net.connected()) {
    connect();
  }
  // send a message every second
  if (millis() - lastMillis > 1000) {
    lastMillis = millis();
    client.publish("/tims-channel", "hello");
    lastMillis = millis();
  }
}
```

4. Copy the contents of the `loop` function to the end of the `loop` function of our main sketch, after the serial code.

The `loop` function in our main sketch should look like this now:

```
void loop() {
  if(Serial.available() > 0) {
    String text = Serial.readStringUntil('\n');
    Serial.print("Received: "); Serial.println(text);
    setText(text);
  }
  client.loop();

  if (!net.connected()) {
    connect();
  }
  // send a message every second
  if (millis() - lastMillis > 1000) {
    lastMillis = millis();
    client.publish("/tims-channel", "hello");
    lastMillis = millis();
  }
}
```

Now let's have a look at the `setup` function. The `setup` function of the MQTT example looks like this:

```
void setup() {
  Serial.begin(115200);
  connect();
}
```

If you compare it to the `setup` function of our main sketch, you will see that we already have the `Serial.begin(115200);` line, and therefore we only need to integrate the call to `connect`.

5. Copy the `connect();` line to the end of the `setup` function of our main sketch:

```
void setup() {
    Serial.begin(115200);
    Serial.println("starting...");
    display.init(115200);
    setText("Hello display");
    display.powerOff();
    connect();
}
```

6. Upload the code and open the Arduino serial monitor. By looking at the display, we can see that it is still working. So far so good.

Open the Arduino serial monitor and you should see incoming data from the MQTT server (messages that we send ourselves every second):

```
incoming: /tims-channel - hello
incoming: /tims-channel - hello
incoming: /tims-channel - hello
...
```

So we know the display is working together with MQTT. Great!

We now need to do the following:

- Delete the messages we send ourselves
- Define a better channel name
- Call our `setText` function to display the incoming text from the MQTT server
- Change our channel-subscription to QoS1 to make sure messages are delivered reliably and nothing is lost

Let's first get rid of the code that automatically sends MQTT messages.

7. In your `loop` function, delete the following lines:

```
// send a message every second
if (millis() - lastMillis > 1000) {
    lastMillis = millis();
    client.publish("/tims-channel", "hello");
    lastMillis = millis();
}
```

8. Scroll to the end of the `connect` function. In the last line, we subscribe to the `/tims-channel` MQTT channel :

```
client.subscribe("/tims-channel");
```

As we want to bundle all our MQTT communication from various devices in the same channel, we create a sub-channel by appending `/todo-device` to it. This represents all data concerning our to-do-device. Because we might add more data points later on, we can be even more specific. We want text that is published to the `client.subscribe` channel via `/tims-channel/todo-device/text` to be shown on the display, so we need to subscribe to it explicitly.

9. Change your subscription to the following:

```
client.subscribe("/tims-channel/todo-device/text");
```

Feel free to use your own name here, so for example, `/marias-channel/todo-device/text`.

Remember topic wildcards, introduced in `Chapter 3`, *Getting Started with MQTT*? We could also subscribe to all messages published to `/tims-channel` like this:

```
client.subscribe("/tims-channel/#");
```

This would subscribe to all messages sent to `/tims-channel` or any of its sub-channels. This might be problematic though. If we add a lot of devices to our fleet of IoT devices, we would get the messages of all of them. This is not what we want.

A saner way using wildcards would be to subscribe to all topics regarding the `todo-device`:

```
client.subscribe("/tims-channel/todo-device/#");
```

But we don't need this right now, and stick with our strict channel subscription, which is specifically for the `text` sub-channel:

```
client.subscribe("/tims-channel/todo-device/text");
```

We can, later on, add more subscriptions if we need to.

Before we move on, there is one more thing that we should do. Per default, subscriptions are using **Quality of Service 0** (**QoS 0**). We talked about this in `Chapter 3`, *Getting Started with MQTT*, and it is an elemental part of MQTT. QoS 0 basically means *I want to subscribe to XYZ, but I don't care if a message might get lost on the way*. If you are sending sensor data via MQTT, this is probably a good choice as it is fine if a message gets lost in between every now and then. QoS 0 is the most lightweight of the QoS settings, but in our case, not the best choice. The alternatives are QoS 1 (at least once) or QoS 2 (exactly once). QoS 2 is heavier because more packages have to be sent for making sure a message has been delivered and it was for sure not delivered more than once.

When we send messages to our to-do device, it would not be a big problem if a message was delivered twice. Adding a check to not update the display with the same text is easy to add. Between possible lost messages using QoS 0, a lot of overhead with QoS 2, you find the sweet spot for our project in the middle: messages sent via QoS 1 will be delivered for sure but don't add too much overhead.

Let's have a look at the documentation of the Arduino MQTT library to see how we can use QoS 1.

10. Head over to the official repository page of the library: `https://github.com/256dpi/arduino-mqtt`.

11. After scrolling down a bit, you will see all possible ways how to use the `subscribe` function:

```
bool subscribe(const String &topic):
bool subscribe(const String &topic, int qos);
bool subscribe(const char topic[]);
bool subscribe(const char topic[], int qos);
```

So there are two ways on how to call it: with QoS as the second parameter, or without. These two versions can be used either with a topic name of type string or type char array. We talked briefly about this before. Some functions only accept one of the two, char array or string. In this case, both are supported. In general, using the `String` class as we do is a little bit easier.

Okay, now we know that we can simply add another parameter to our existing call to `subscribe`, passing 1 as QoS.

12. Change your `client.subscribe` call to the following:

```
client.subscribe("/tims-channel/todo-device/text", 1);
```

By specifying the QoS as 1 on the channel subscription, we make sure that messages that already reached the MQTT server will be delivered to us. If on the way there was an error, the MQTT server will try again and again, until it is delivered.

But this only affects the connection between the Arduino and the MQTT server. If the device sending a message to the channel we subscribed to is using QoS 0 (fire and forget), we cannot be sure that we will receive it. Therefore, both ends have to use QoS 1 (or QoS 2). We will see how to send messages using QoS 1 later.

We are done with our subscription now. Messages sent to the `/tims-channel/todo-device/text` channel will be delivered to the Arduino; we just don't do anything with the incoming text.

The function that handles incoming messages is `messageReceived`. It has two parameters, one for the topic name and one for the payload—the data. In our case, `payload` will contain the text to be displayed:

```
void messageReceived(String &topic, String &payload) {
    Serial.println("incoming: " + topic + " - " + payload);
}
```

To display the incoming text on our display, all we have to do is add a call to our `setText` function, passing it the incoming text.

13. Change your `messageReceived` function to the following and re-upload the sketch:

```
void messageReceived(String &topic, String &payload) {
    Serial.println("incoming: " + topic + " - " + payload);
    setText(payload);
}
```

That's it. Now your device is ready to display incoming MQTT messages. In the next section, we will send MQTT messages to the channel we subscribed to and find out if they are successfully shown on the display.

You can find the code until here as a checkpoint called `ch6_03_mqtt` in the downloaded source code repository in the `ch6 | arduino` folder.

Sending messages via MQTT

Our device is now ready to receive messages via MQTT. The MQTT server we are using features an MQTT, as well as an HTTP-interface. There are many different ways that we can now send messages to our device:

- Via Terminal/PowerShell as MQTT messages using an MQTT client such as Mosquitto
- Via Terminal/PowerShell as HTTP messages using `curl`
- Via a third-party iOS/Android app
- Via a third-party macOS/Windows/Linux app
- Via a third-party web-based client (for example, `http://www.hivemq.com/demos/websocket-client/`)
- Via a custom-made website using a JavaScript-MQTT client
- Via a custom Android/iOS/macOS/Linux/Windows app using MQTT-client

Basically everything is possible depending on how you send text to your device. MQTT is amazing!

Because programming a custom app would be a bit much on top of what you have already learned, we will focus on the easier ways to send MQTT messages. The following ways are very easy to use and are open for you to use whether you are a beginner or intermediate programmer:

- Via Terminal/PowerShell as MQTT messages using an MQTT client such as Mosquitto
- Via Terminal/PowerShell as HTTP messages using `curl`
- Via third-party iOS/Android app
- Via third-party macOS/Windows/Linux app
- Via third-party web-based client (for example `http://www.hivemq.com/demos/websocket-client/`)

They all work in the same way. You have to provide the MQTT server, port, username, and password, specify what you want to publish (payload), where you want to publish it to (topic), and which QoS setting you want to use (optional, mostly defaulting to QoS 0).

Feel free to experiment especially with the MQTT apps and web-based websites presented in Chapter 3, *Getting Started with MQTT*. They make it really easy to send MQTT messages out of the box without much setup.

Enough talking, let's send some messages to be displayed on your device. We will start sending MQTT messages using the Terminal (macOS) or PowerShell (Windows). I will refer to Terminal for both variants. If you are using Windows, substitute Terminal with PowerShell. The commands to be executed are exactly the same.

In Chapter 4, *Setting Up a Lab Environment*, we made sure you set up Mosquitto, a free MQTT server and client, to be used with the Terminal.

Please follow along with these steps:

1. Make sure Mosquitto is working correctly by running the following command in Terminal:

    ```
    mosquitto_pub
    ```

    If it was installed correctly, you should see its help output (which is common for command-line tools when no arguments were given). The first lines of the output should look like this:

    ```
    mosquitto_pub {[-h host] [-p port] [-u username [-P password]]
    -t topic | -L URL}
                    {-f file | -l | -n | -m message}
                    [-c] [-k keepalive] [-q qos] [-r]
                    [-A bind_address]
                    ...
    ```

2. Have a look at all the available options. Hopefully, you recognize a lot of them from what we discussed in Chapter 3, *Getting Started with MQTT*.

    We need to specify the following parameters, which are listed in the help output:

    - -h: MQTT host to connect to. Defaults to localhost.
    - -u: Provide a username.
    - -P: Provide a password (please note that this is an uppercase P!).
    - -t: MQTT topic to publish to.
    - -m: Message payload to send.

- -q: Quality of service level to use for all messages. Defaults to 0.
- -p: Network port to connect to. Defaults to 1883 for plain MQTT and 8883 for MQTT over TLS.
- -r: Message should be retained.

Let's start to collect the information we need to construct an MQTT publish command.

3. Visit the shiftr.io MQTT interface documentation on the shiftr.io website (https://docs.shiftr.io/interfaces/mqtt/).

Here, you will find most of the information we need in order to use Mosquitto with shiftr.io.

You can see the Mosquitto parameters together with the matching information from shiftr.io, as follows:

- -h: broker.shiftr.io (MQTT server)
- -p: 1883 (Port, as 1883 is the default port, we don't have to specify it explicitly)

Without creating an account on shiftr.io, you can use the open login credentials that can be used by everyone:

- -u: try (MQTT username)
- -P: try (MQTT password, uppercase P)

Feel free to use your own login credentials later on.

4. Enter the following into your Terminal and press *Enter*:

```
mosquitto_pub -h broker.shiftr.io -u try -P try -t "/tims-
channel/todo-device/text" -m "Hello hello"
```

Tadaa! Your display will refresh and show the text Hello hello:

Text Hello hello shown on the display, sent via Mosquitto MQTT client

Now let's add some more options to it. To make sure that our message is delivered correctly, we have to send the message using QoS 1 or QoS 2. As previously discussed, using QoS 2 is overkill for our use-case and using QoS 1, in this case, is the best choice. Using QoS 1, messages will be delivered for sure, but they might be delivered more than once, which is not a problem for us in this case.

5. Inside your Terminal, press the up key to re-use the command that you just entered.
6. Append `-q 1` to it to send the message with QoS 1 instead of QoS 0 (default). The command should look like this now:

```
mosquitto_pub -h broker.shiftr.io -u try -P try -t "/tims-
channel/todo-device/text" -m "Hello with QoS 1" -q 1
```

7. Press *Enter* to execute the command and you should see `"Hello with QoS 1"` on your e-paper display.
8. Now disconnect your Arduino and re-connect it again.

The display will refresh and show `"Hello display"` because of the `setText("Hello display");` line in our `setup` function, but it does not show the last text we sent to our MQTT channel (`"Hello with QoS 1"`). Not very usable as a to-do list. We can easily fix this using retained messages. Do you remember retained messages from `Chapter 3`, *Getting Started with MQTT*? Retained messages are buffered per topic, so the last message sent to a channel (with the retained flag set) will be kept and sent to new subscribers. So even when the Arduino was disconnected, when it re-connects, the MQTT server will send the last retained message to it, kind of like a little database. Each channel can hold exactly one element.

9. Disconnect your Arduino again, then publish a message with the retained flag (–r):

```
mosquitto_pub -h broker.shiftr.io -u try -P try -t "/tims-
channel/todo-device/text" -m "Hello Retained" -q 1 -r
```

10. Re-connect your Arduino, and after a few seconds, your display will show `Hello Retained`:

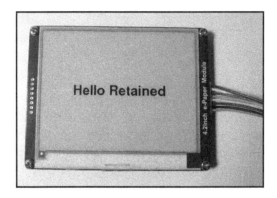

The retained message was received even while the Arduino was offline when the message was sent

Great, our mini MQTT-based database is working!

As of writing this, the Arduino MQTT library is not fully optimized for the Arduino MKR WiFi 1010, which might lead to connection problems between the Arduino and the MQTT server. If you look in the Arduino serial monitor, you might notice some disconnects followed by reconnects. In our code, we check in the `loop` function if the Arduino is connected to the MQTT server. If it is not, we simply reconnect. While this makes sure that we are not disconnected for too long, we receive the same retained message over and over (every time the connect method is called) and therefore, we resubscribe to the `/tims-channel/todo-device/text` topic.

The result is that the display keeps refreshing. We can easily fix this by storing the text we are about to display and before we set it the next time, check whether it is different text:

1. Create a new `lastText` variable in the top part of your sketch, just before the `setup` function:

```
unsigned long lastMillis = 0;
String lastText = "";
```

2. Add the following `if` condition in our `setText` function to check whether the text we are about to display is already displayed:

```
void setText(String text) {
    if (text.equals(lastText)) {
        return;
    }
    display.setRotation(0);
    ...
```

We use the `equals` function of the `String` class to compare the text we are about to display with our newly created `lastText` variable. Now the only thing left to do is update `lastText` after we ran the code to update the display panel.

3. On the bottom of the `setText` function, add the following line:

```
        ...
        display.print(text);
    } while (display.nextPage());
    lastText = text;
}
```

4. Re-upload the code and you will see that the screen will display the last retained value `Hello Retained`, but it has stopped flashing wildly.

5. Now try to send a message to be shown as important on the display (for example, `Super Important!`) by adding an exclamation mark as the last character:

```
mosquitto_pub -h broker.shiftr.io -u try -P try -t "/tims-
channel/todo-device/text" -m "Super Important!" -q 1 -r
```

Sadly this might not work, depending on your environment. It is very likely that you see the following error message:

```
-bash: !": event not found
```

Using an exclamation mark in the string might lead to an error if your Terminal application uses Bash. If you are using macOS or Linux, it is very likely that you will see the previous error, which is caused because `!` directly followed by a `"` character has a special meaning. But there is an easy fix for this problem. Instead of using double quotes, we can use single quotes.

6. Change the `mosquitto_pub` command to the following:

```
mosquitto_pub -h broker.shiftr.io -u try -P try -t "/tims-channel/todo-device/text" -m 'Super Important!' -q 1 -r
```

Instead of writing `"Super Important!"`, we simply write `'Super Important!'` (using single quotes) to overcome the problem we had before.

Please note that you could also put quotes around the username and password. If either of them use special characters, for example, a space, you definitely must use quotes around them, otherwise it will lead to an error. But we can just leave it like it is, because `try` does not use any special characters:

Display with text Super Important! being shown, sent from via the Mosquitto client

I don't want to go into too much detail here on how using single quotes, double quotes, and no quotes differ in the Terminal. As a rule of thumb, you can follow these guidelines:

- You don't need to use quotation marks when your string is just a simple word without special characters or white space, for example, `try`.
- In most cases, using double quotes is recommended, for example, `"/tims-channel/sub-channel with a space"`.
- In some cases, you need to use single quotes to prevent your Terminal from interpreting the string. In our case, we have to use single quotes for our message because ! is interpreted in a special way. Writing `'Hello!'` instead of `"Hello!"` fixes this.

While this might sound very complicated (and indeed it is, there are many rules behind it), you don't need to think too much about it. I just want you to know that there is a difference between using no quotes, double quotes, and single quotes when passing strings to commands in the Terminal. If you run into any problems like the previous one, I hope you will remember: *wait a minute, wasn't there that thing with the quotes?*

Enhancements and building a case

Right now, your prototype looks very much like a prototype—an e-panel module connected with a bunch of wires to an Arduino MKR WiFi 1010. In Chapter 8, *Presenting Your Own Prototype*, you will get an idea of how to present it in a nicer way. The first step, which does not consume too much time, is to create a case using thick paper or cardboard. Typically you would measure the dimensions of your components and build a custom case for it using paper/cardboard, a pencil, tape, and scissors.

Do you have some ideas about what modules/components you could add to it? Maybe a button to reset the text? When your e-paper to-do device reminds you to take out the trash and you prepare to leave the flat, it would be nice to just press a physical button to clear the display instead of getting out your smartphone and sending an MQTT message to clear the display, wouldn't it?

All you need to do is connect an additional button to your Arduino MKR WiFi 1010, and once it is pressed, it will publish a new (empty) message to clear the screen. For this, you would first need to find out what options the Arduino MQTT library offers for publishing messages. Looking at the GitHub page of the library (https://github.com/256dpi/arduino-mqtt), you can see the following:

```
bool publish(const String &topic);
bool publish(const char topic[]);
bool publish(const String &topic, const String &payload);
bool publish(const String &topic, const String &payload, bool retained, int
qos);
bool publish(const char topic[], const String &payload);
bool publish(const char topic[], const String &payload, bool retained, int
qos);
bool publish(const char topic[], const char payload[]);
bool publish(const char topic[], const char payload[], bool retained, int
qos);
bool publish(const char topic[], const char payload[], int length);
bool publish(const char topic[], const char payload[], int length, bool
retained, int qos);
```

The fourth one looks perfect for us. Similar to the command sent via the Terminal, using Mosquitto, we can send it as a retained message and also specify the QoS level:

```
bool publish(const String &topic, const String &payload, bool retained, int qos);
```

So a call to reset the display would look like this:

```
publish("/tims-channel/todo-device/text", "", true, 1);
```

Then, your `publish` command could look like this:

```
publish("/tims-channel/todo-device/status", "Online", true, 1);
```

I hope you see this project as a starting point for your own ideas. There is a lot that can be done using an internet-connected e-paper module.

You can see my finished display in the following photograph. Using the beginner friendly 3D design tool Tinkercad (`https://www.tinkercad.com`), I designed a case for the prototype. In the bottom center, I also included a button to reset the display.

You can find the 3D design files in the book's repository in the `ch6 | 3d` folder, so if you have access to a 3D printer, you can print it out yourself. In `Chapter 7`, *Building a Smart Productivity Cube, Part 1*, I will give you an introduction to Tinkercad, so you will learn how to build cases like this yourself. It is easier than you might think.

Finished prototype with a custom case and an additional clear button, the software Tinkercad

In the book's repository, you will also find the code for another way to send text to your prototype: a custom-made website with a text input that allows you to send MQTT messages to the MQTT server via a little bit of JavaScript. Due to the scope of this book, we cannot get into too much detail about it here, but feel free to check it out. You can find it in the `ch6/web` folder. Have a look at the `README.md` file (open it with any text editor, or even better, view it with a markdown editor).

The HTML, CSS, and JavaScript files can be opened using Visual Studio Code, which we installed together in `Chapter 4`, *Setting Up a Lab Environment*.

Summary

In this chapter, you learned how to display text on an e-paper display and how to send text to it either via serial port or MQTT.

You learned how to make use of MQTT features such as QoS 1 and retained messages. QoS 1 makes sure messages are delivered at least once. The last message sent with the retained message flag is stored for each channel. When clients subscribe to it, they will instantly receive the last available retained message, even when no message has been published since the Arduino resubscribed.

You also learned how to send messages using different QoS settings as well as the retained message flag via the free command-line tool Mosquitto.

In the next chapter, you will build another MQTT-connected device: a smart productivity cube. Using a custom made orientation sensor, this device can be used to track the time that you spend watching TV or exercising, for example. We will also learn how to use MQTT client apps for Android and iOS to receive the information published by your smart device.

Questions

1. What does QoS stand for?
2. Why are we using QoS 1 instead of QoS 0 or QoS 2?
3. How can messages be sent to the MQTT server besides using the Mosquitto command-line tool?
4. What is a client ID?
5. Can there be multiple `loop` and `setup` functions in one sketch?
6. Try using one of the Android or iOS app presented in `Chapter 3`, *Getting Started with MQTT*, to publish messages to our channel.
7. Try using your own login data of the shiftr.io server instead of the public `try`/`try` login.

Further reading

- **Waveshare e-paper module product information website**: `https://www.waveshare.com/wiki/4.2inch_e-Paper_Module`
- **shiftr.io MQTT reference**: `https://docs.shiftr.io/interfaces/mqtt/`

7
Building a Smart Productivity Cube, Part 1

In the previous chapters, we built a smart pet food dispenser and a smart e-ink to-do list. In this chapter, we are going to build a smart productivity cube. By using mechanical tilt switches, we will be able to sense which side the cube is standing on. We will then assign that position to an activity (for example, learn MQTT) and log how long we pursue this activity.

After following along with this chapter, you will know how to build your own orientation sensor using tilt switches. This chapter serves as the basis for the next chapter, Chapter 8, *Building a Smart Productivity Cube, Part 2*, where we will learn how to enhance your smart productivity cube with MQTT and how to display its information on your smartphone using Android and iOS apps.

In order to facilitate the learning process, the chapter is divided into the following sections:

- Building the smart productivity cube
- Building the cube
- Detecting orientation changes

Technical requirements

Before we begin, please make sure that you read Chapter 4, *Setting Up a Lab Environment*. In Chapter 5, *Building Your Own Automatic Pet Food Dispenser*, we also tested the required MQTT and Wi-Fi libraries (in the *Testing MQTT on Arduino* section).

Besides the MQTT and WiFiNINA libraries, we do not need any additional libraries for the Arduino IDE.

The following components are needed to build the project:

- **Arduino MKR WiFi 1010**: The development board we will be using: `https://store.arduino.cc/usa/mkr-wifi-1010`
- **Micro USB cable**: To connect the Arduino to your computer: `https://www.sparkfun.com/products/10215`
- **Jumper wires (female-male, soft)**: To connect the tilt switches to the Arduino: `https://www.sparkfun.com/products/12794`
- **4 tilt switches**: To detect device orientation: `https://www.sparkfun.com/products/10289`
- **Jumper wires (hard)**: To increase the stability of our prototype: `https://www.sparkfun.com/products/124`
- **Half-sized breadboard**: To create the circuit on: `https://www.sparkfun.com/products/12002`
- **Scissors, cardboard, tape, glue, and a sponge**: To build a case for our device

You can find the source code for this chapter in the book's repository (`https://github.com/PacktPublishing/Hands-on-IOT-with-MQTT`) in the `ch7` folder.

Check out the following video to see the Code in Action: `http://bit.ly/2oSjufZ`

Building the smart productivity cube

Before we get started building the smart productivity cube, let me first explain what exactly we are going to build and why.

Many people seem to suffer from the same problem: they feel like they have too little time and wonder where their time went. There are already plenty of time-tracking applications for desktop computers, smartphones, and tablets, such as Toggl (`https://toggl.com`) and Harvest (`https://www.getharvest.com`), that address this problem. All of them serve the same purpose—counting how much time you spend on what activities, so you have a better understanding of how you use your time. This is either used for personal insights or professionally, for example, to count billable hours when working for a client. Most of them work like a stopwatch does—you manually start and stop a timer, one per task.

While most of these apps work just fine, they all have a major downside: you have to open the tracker app (on a computer, smartphone, or tablet) in order to see whether the tracker is currently on and what task it is tracking.

We will fix this by building a physical device that will serve the same purpose: tracking time for various things. Instead of pressing a button in an app, though, we will physically rotate a cube. This way, we can use it also when our computer and smartphone are shut down. Also, by looking at it, we will always see what it is currently tracking—no need to open an app.

Now, you might ask yourself how we will be able to see how much time we spend on the tasks that we are about to track if there is no screen involved.

Well, of course, we are using MQTT for tracking. After building the physical tracking device, we will build an interface using one of the available MQTT apps. They make it really easy to display the information sent to a specific channel and can, therefore, be used to display the tracked time.

To build a smart productivity cube, very few electrical components are needed.

We need a mechanism to find out which side our cube is standing on. For this, we have two options:

- The first option is to use a gyroscope (provided for example by the SparkFun IMU Breakout—MPU-9250). Gyroscopes are often used in robotic applications to sense the angle of a device. They are great and there are many use cases for gyroscopes, but they don't perform well for 360º tracking:

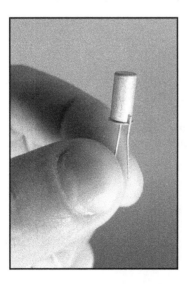

A tilt switch: inside the can is a tiny metal ball, which closes the connection between the two legs when held upright

- The second option for orientation tracking is using tilt switches. These tiny components behave very similarly to a push button—either the connection is closed or not. While with a push button the person pressing it is closing the circuit, using a tilt switch, gravity closes it (when the tilt switch is rotated). Inside the can is a tiny metal ball. When the tilt switch is held upright (with the legs pointing down), the metal ball closes the connection between the two legs; when it is turned around (with the legs pointing up), the ball falls to the other side and the connection is open.

There are a lot of tutorials found online or in books that involve the use of a button. In each and every one of them, you can replace the push button with a tilt switch and the code will still work. People trying out a prototype are often amazed when a form of interaction is turning the object upside down, instead of just pressing a button. Push buttons are everywhere and we are used to pressing them every day (on TV remotes, elevators, and microwaves, for instance); turning objects upside down to activate a function is rarely seen in everyday life.

In our project, we are going to use four tilt switches to sense the cube on all six sides. We could also create other shapes with even more sides, but then it gets a little bit tricky because we would need more tilt switches and we would need to arrange them very precisely.

In our case, to sense six sides, four tilt switches are enough.

Using one tilt switch

Let's start by attaching one of the tilt switches. Using one instead of all four at the same time makes it easier to verify that it is working and to understand how it works:

1. Connect one of the tilt switches in the same way as in the following diagram—one pin connected to ground, the other to pin **0**:

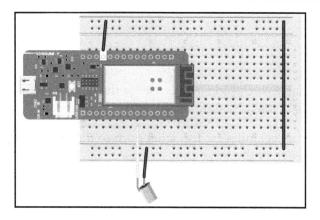

Arduino MKR WiFi 1010 connected to a tilt switch (this image was created with Fritzing)
License: (https://creativecommons.org/licenses/by-sa/3.0/)

In reality, this will look something like this:

Arduino MKR WiFi 1010 and tilt switch (connected to ground and pin 0)

2. Now upload the following code:

```
int SWITCH_PIN = 0;

void setup() {
  Serial.begin(115200);
  pinMode(SWITCH_PIN, INPUT_PULLUP);
}
```

```
void loop() {
  bool s1 = digitalRead(SWITCH_PIN);
  Serial.println(s1);
}
```

First, we declare which pin we will use for the tilt switch (in our case, 0). We then do two things in the `setup` function:

1. First, we initialize the serial port, so that we can log the value of our switch to the serial port. We use 115200 as the baud rate here, which allows for fast communication.
2. In the second line of the `setup` function, we define SWITCH_PIN as input, in the same way as we would do for a push button. The second parameter to the `pinMode` function is important here. If you have worked with buttons on the Arduino before, you probably know about pull-up and pull-down resistors.

Whenever you use a button, or a button-like component (in this case, a tilt switch), in your project, you need to make sure that the input pin is never floating. The input value can be floating (switching from 0 to 1 frequently) when there is no connection on the pin. In our case, this is true when the button is not pressed.

See how the output changes when you move the construction with the tilt switches around (by rotating it 90° or 180°):

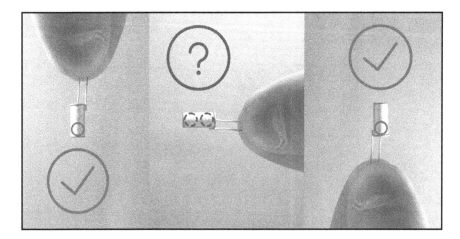

Tilt switch positions: the flat position cannot be read reliably

Tilt switches can be read reliably when the little metal ball inside is in a stable position. The tilt switch in the center photo cannot be read reliably. Its value could be either 1 or 0. So, tilt switches arranged at a 90° angle should not be used.

By activating the pull-up resistor on the pin using `pinMode(SWITCH_PIN, INPUT_PULLUP)`, internally, the pin is pulled toward the Arduino's 3.3V power source using a high internal resistor. So, when the pin has no connection, internally, the current is flowing anyway. The pin is pulled toward 3.3V and its value is `HIGH`. This means the signal is inverted. Due to the high impedance of the internal resistor, the current only flows that way if the button is not pressed (or, in our case, the connection of the tilt switch is open). If the button is pressed (or the tilt switch connection is closed), the pin is connected to ground. Its input will then be `LOW` (because it is inverted).

If you are not familiar with floating states, you should watch the video *Button Tutorial: Arduino Built-In Pull-Up* by James Lewis on YouTube (`https://www.youtube.com/watch?v=jJnD6LdGmUo`).

There are a few more lines of code we need to talk about:

```
void loop() {
  bool s1 = digitalRead(SWITCH_PIN);
  Serial.println(s1);
}
```

Inside the `loop` function, we do two things. First, we read the value of our tilt switch pin and store it in a variable called `s1`. This variable is of type `bool`, because the value returned by `digitalRead` is either 0 (`LOW`) or 1 (`HIGH`). Because we activated the internal pull-up resistor, which prevents a floating state, the value of `s1` will be inverted, so it will be 0 when the tilt switch is connected upright (legs down) and 1 when it is held upside down (legs up).

3. Now open the Arduino serial monitor by clicking on **Tools** | **Serial Monitor** and you should see a lot of lines being printed, either 0 or 1, depending on how you hold the tilt switch. Take some time to get to know how it reacts. At what angle is the ball moving (and therefore the state switching)? By playing around with it, you should get a feeling of how accurate it is and maybe get some ideas on what other projects you might be able to use it for.

Let's now connect the other switches.

Connecting the other tilt switches

By now, you should have a feeling for how the tilt switch behaves and what values it outputs when you hold it upright or upside down.

With one tilt switch, we are able to sense two positions, either upright or upside down. We want to sense even more positions by using multiple tilt switches. If we arrange them at different angles, we can reliably sense various orientations.

Because we want to build a smart productivity cube, we need to sense six sides: bottom, top, left, right, front, and back. We need to lay out the tilt switches in such a way that all of these sides can be reliably sensed. When I say reliably, I mean that when placing the cube on one of its sides, the output of the tilt switches should be stable (not changing), so we know which side it is standing on.

The first idea when using tilt switches to sense all sides of a cube might be to use one for each of the axes: one for the x axis, one for the y axis, and one for the z axis. The problem with this approach is that it would produce a lot of unpredictable behavior. When the tilt switches are arranged at 90° angles, the ball might sometimes roll to one side or the other. We would have a floating state similar to the one described before with the raw pin readings, but this time caused by the small metal ball inside the tilt switch—sometimes it might move to one side, sometimes to the other. This is not reliable at all; we need a better way.

When using four tilt switches and arranging them at 45° angles instead of 90° ones, we will be able to sense all six sides without the floating states. When the cube is placed on either of its six sides, the metal balls in the tilt switches will always have a defined state. This solves our problem. We can then sense all six sides and have a reliable signal.

Let's make it happen!

1. Connect the other three tilt switches so that you have a total of four tilt switches connected. It is good to always use the same color for everything that is connected to ground (black or blue) and an individual color for each cable going from one of the tilt switch pins to the digital pins of the Arduino. Using different colors will make it easier later on to identify the individual tilt switches:

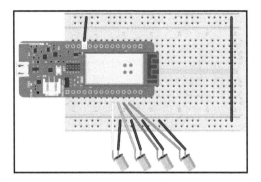

Arduino MKR WiFi 1010 connected to four tilt switches (this image was created with Fritzing)
License: (https://creativecommons.org/licenses/by-sa/3.0/)

In the following photo, you can see how it looks when the four tilt switches are connected to the Arduino MKR WiFi 1010:

Four tilt switches connected to the Arduino. The orange cable connects both ground terminals (blue).

2. Now we need to adapt our code a bit to use the four tilt switches instead of just the one. You can also find this code in the book's repository (`https://github.com/PacktPublishing/Hands-on-IOT-with-MQTT`) in the `ch7_02_four_switches` folder.
3. Change your code to the following:
    1. First configure the variable definitions:

```
int SWITCH_PIN1 = 0;
int SWITCH_PIN2 = 1;
int SWITCH_PIN3 = 2;
int SWITCH_PIN4 = 3;
```

2. Then configure the `setup` function:

```
void setup() {
  Serial.begin(115200);
  pinMode(SWITCH_PIN1, INPUT_PULLUP);
  pinMode(SWITCH_PIN2, INPUT_PULLUP);
  pinMode(SWITCH_PIN3, INPUT_PULLUP);
  pinMode(SWITCH_PIN4, INPUT_PULLUP);
}
```

3. Follow that with the `loop` function:

```
void loop() {
  bool s1 = digitalRead(SWITCH_PIN1);
  bool s2 = digitalRead(SWITCH_PIN2);
  bool s3 = digitalRead(SWITCH_PIN3);
  bool s4 = digitalRead(SWITCH_PIN4);
  Serial.print(s1);
  Serial.print(", ");
  Serial.print(s2);
  Serial.print(", ");
  Serial.print(s3);
  Serial.print(", ");
  Serial.print(s4);
  Serial.println(", ");
}
```

4. Upload the code now and, after opening the Arduino serial monitor, you will see one sensor reading being printed instead of four. If you lift one of the tilt switches up, you should see one of the numbers change. But we will look into that in more detail later on.

To use the four tilt switches as a six-sided sensor, we need to arrange the tilt switches first. Follow these steps:

1. Cut two thin cardboard strips (8 × 1 cm):

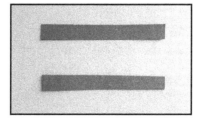

Cardboard strips

2. Align them to form an X and tape them together:

Cardboard strips forming an X

3. Tape the tilt switches to the cardboard (cables pointing inward). If your wires are as short as mine, it might be easier to disconnect the cables from the Arduino (as I did) and reconnect them afterward:

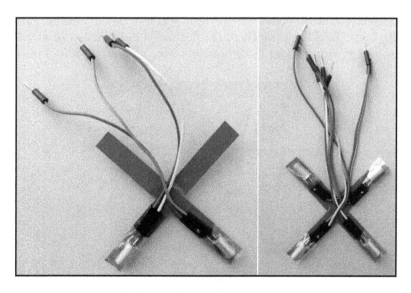

Tilt switches connected to the cardboard strips with tape

4. Now we need to fold the cardboard on each side where the two strips meet in the middle so we can form a pyramid by lifting the middle part up:

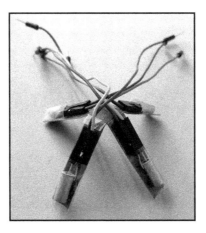

Elevated tilt switch construction

5. Next, reconnect the cables. It is advisable to connect them in the same order as I did in the following photo—so, bottom tilt switch = pin **1**, right pin = pin **0**, top pin = pin **2**, and left pin = pin **3**. The other leg of each tilt switch must be connected to ground (**GND**). If you connect the pins in another order, this is totally fine. The only thing that is important is that one leg of each tilt switch is connected to ground and the other leg to one of the **0**, **1**, **2**, or **3** pins:

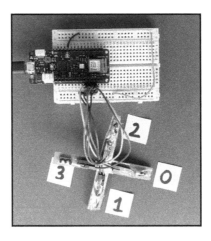

Pin numbers show the digital pins on the Arduino being used

6. The last step for creating the base for our smart productivity cube is to tape our tilt switch X to the breadboard. When doing so, make sure that the center of our X is elevated, forming a pyramid:

Elevated tilt switch construction on the breadboard

Here you can see the tilt switch construction from above:

Prototype from above

Great! We are all set now. Don't worry if it looks a bit hacky. We will, later on, build a cube around it to make it look nicer.

If you haven't done so already, upload the code. Then open the Arduino serial monitor.

You will notice that now four 0 values and four 1 values are being printed—the (inverted) states of the four tilt switches.

Great—you just created a 3D orientation sensor from scratch! Try to imagine a cube around your construction and move it in 90° steps in your hands. Have a look at the serial output. Is it consistent? Or does one of the tilt switches need to be readjusted to not get into a floating state where the ball is not in a stable position when rotated in any 90° direction (imagine the cube we are about to build later around our breadboard to simulate all six positions)?

Adjusting the construction should not take too long. Because we are only sensing six sides, we have a lot of tolerance for how exact the angles need to be in order to get accurate readings. The only angles that we cannot use are in the range between 80° and 110°. In our case, the angles are between 30° and 70°, so we should not have a problem with floating states (only temporarily when moving the cube from one side to another, but we will come to that later).

If we were to build a polygon with 20 sides as in the dice of the famous role-playing game Dungeons & Dragons, it would be a lot harder to find a layout for the tilt switches to reliably sense any of the 20 sides, because you would have to make sure that none of the tilt switches were arranged in a way that the value was floating. Making sure that none of the pins is floating on one side might be doable here, but doing so for 20 sides is quite a challenge.

We now need to find out which combination of values of the tilt switches corresponds to each of the six sides. When the construction is standing on your table, the values of all tilt switches should be 1. The little metal balls inside the tilt switches are not closing the connection. But since we are using tilt switches, closing the connection using internal pull-up resistors results in the value being 0. If you have problems understanding how pull-up resistors work and why the value is 0 when the connection is closed (instead of 1), don't worry. I would recommend watching the video mentioned before to get a better understanding of it, but to proceed with this project, you just need to understand that combinations of tilt switch values correspond to one of the sides. 1, 1, 1, 1, for example, is printed when the construction is just lying on the table (default position).

The code until here can be found in the repository as ch7_02_four_switches.

Detecting the sides

Let's create the mappings to have a more readable output. Instead of, for example, 1, 1, 1, 1, we want to have an output such as `Standing on: bottom` or `Standing on: left side`. Follow the steps to add side detection to your cube:

1. Rotate the construction for each side and write down what the serial monitor prints on a piece of paper. My mapping looks like this:

   ```
   1 1 1 1: Side 0 (Bottom)
   0 0 0 0: Side 1 (Top)
   0 1 0 1: Side 2
   1 0 1 0: Side 3
   1 1 0 0: Side 4
   0 0 1 1: Side 5
   ```

   By looking at the number combinations, you can see that either all pins are 1 (bottom side), all pins are 0 (top side), or two pins are 0 and two pins are 1. Each side has a different combination of tilt switch readings.

2. Before we add the rules to detect the preceding six cases mentioned, we need to create two variables. Put the following code at the top of your sketch, right before the `setup` function:

   ```
   int previousSide = -1;
   int currentSide = -1;
   ```

   It declares a variable to hold the ID (identifier) of the side currently facing down. −1 is used as a starting value to indicate that it does not hold an actual side ID, yet.

   We will assign an ID to each side, a number between 0 and 5.

   While it might be more convenient to use a string as an ID for each individual side (top, bottom, left, right, front, back, for instance), that makes an assumption regarding how you want to place the cube. Maybe you want to use it differently, as I am. Having a mis-mapping, for example, if the ID "bottom" meant "left" in your case, would be very confusing.

The other reason is that comparing integers is much more performant than comparing strings. To detect whether the cube has been flipped, we need to store the previous side it was standing on and compare it to the current one, and do so for each loop iteration. If we could do this comparison less often, having a string-comparison overhead would be more tolerable.

3. Now let's add the rules to detect the different sides. Put the following code at the end of your `loop` function:

```
if (s1 == 1 && s2 == 1 && s3 == 1 && s4 == 1) {
  currentSide = 0; // bottom
} else if (s1 == 0 && s2 == 0 && s3 == 0 && s4 == 0) {
  currentSide = 1; // top
} else if (s1 == 0 && s2 == 1 && s3 == 0 && s4 == 1) {
  currentSide = 2;
} else if (s1 == 1 && s2 == 0 && s3 == 1 && s4 == 0) {
  currentSide = 3;
} else if (s1 == 1 && s2 == 1 && s3 == 0 && s4 == 0) {
  currentSide = 4;
} else if (s1 == 0 && s2 == 0 && s3 == 1 && s4 == 1) {
  currentSide = 5;
}
Serial.print("Side: "); Serial.println(currentSide);
```

4. Afterward, comment the previous print statements in our `loop` function to have a less cluttered serial monitor output:

```
// Serial.print(s1);
// Serial.print(", ");
// Serial.print(s2);
// Serial.print(", ");
// Serial.print(s3);
// Serial.print(", ");
// Serial.print(s4);
// Serial.println(", ");
```

5. Upload the code and see whether it works by looking at the output of the serial monitor and rotating the breadboard construction.

We created a rule for every combination of tilt switch values that was printed to the console earlier. If one of these conditions is met, we update our `currentSide` variable with the specific new ID.

We are one step further now—we are able to sense 90° rotations, the basis for our smart productivity cube.

The code until here can be found in the book's repository as `ch7_03`.

Building the cube

Let's continue by building a cube around the breadboard so that we can place it on different sides. Because we want it to be stable, we need to make sure we secure the breadboard well enough. If we place the cube in a way that the breadboard construction is at the top, we don't want it to fall down.

To compress the size needed on the breadboard, you can move the Arduino a bit inward (placing it on a position on the breadboard so that the Arduino does not stand out). When using a half-sized breadboard, you can then use a side length of 8.5 cm for the cube. If you do so, please don't forget to change the position of the breadboard cables as well. You need to move the cables going from the Arduino to the tilt switches. They need to be connected to ports 0, 1, 2, and 3.

Also, don't forget to move the cable connecting ground to the ground rail, otherwise you will get the same reading on all sides: 1, 1, 1, 1.

To create the base form of the cube, you need to create a scribble of how the layout can be arranged so that it forms a cube in the end that can be closed. If you have large paper, it is probably easiest to make a cube out of cardboard in one piece. If you only have A4 paper, you will need to create separate pieces and combine them. Now that we are all set, follow the following steps to make your prototype look good:

1. In the following photo, you can see my starting point for creating the case. It consists of three sides. One of the sides contains extra areas to glue on. Before starting to build your cube, you should measure how big each side needs to be so that the prototype can fit in:

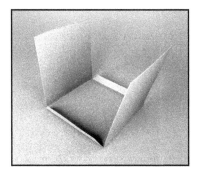

The base for my cube; I used two cardboard parts to form the cube

2. The following photo shows the second part of the cube case. Made out of A4 paper, it consists of three cube sides as well as glue areas. For folding, you can use a hard object, such as a ruler:

To fold the edges, you can use a ruler

3. When you have both pieces ready, add a small hole for the USB cable to fit through on one of the sides, as displayed in the following photo:

These two cardboard parts will form a cube. The part in the back has a cutout for the USB cable.

4. Now it is time to tape the breadboard with the Arduino and tilt switches to the cardboard. When we turn the cube upside down, we want it to stay in place:

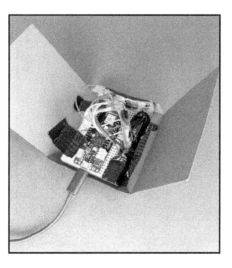

The tape is keeping the breadboard-construction in place when the cube is being rotated

5. See whether your cube is stable enough. Maybe you need to add more tape:

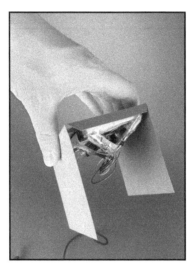

Checking whether the tape is holding the breadboard construction

6. Add some soft material (such as a sponge) on the top to improve stability when the cube is rotated. To prevent the material from touching the pins, you should add a layer of plastic film in between (you can use an old plastic bag here):

Sponge on top of the breadboard construction

7. After the prototype is secured from falling, we can glue all the sides together. Make sure you left a hole for the cable.

You can then put stickers on each side to make clear what activity each side should be tracking. I cut out paper circles for this.

The labels or activity names can be added later as well, once we have written more code:

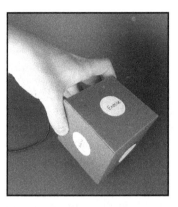

The finished smart productivity cube with activity labels on each side

In the book's repository, you will also find the 3D files for a custom cube case. In `Chapter 9`, *Presenting Your Own Prototype*, I will show you how it has been designed using the free tool Tinkercad (`https://tinkercad.com`):

The productivity cube with a 3D-printed case and colored dots

Now let's add a check to detect orientation changes.

Detecting orientation changes

The easiest approach to detect orientation changes of our cube would be to place the following code at the end of our `loop` function:

```
if (previousSide != currentSide) {
  Serial.print("Side: "); Serial.println(currentSide);
  previousSide = currentSide;
}
```

But the easiest approach is not always the best one. The problem with this approach is that when we switch sides, another side might be detected in between. For example, let's say the cube was lying on the bottom side and you wanted to place it on its top side. While rotating it, it would (for a very short time) detect that it is placed on the side in between, even though we are just rotating it by hand and it does not rest on this side.

To overcome this problem, we should only detect a side change when the last x values were all recorded on the same side. To get this right, we need to follow these steps:

1. Create four new variables and place them below our `SWITCH_PIN` definitions (at the top):

   ```
   int CHECK_MILLIS = 50;
   int CHECK_ITERATIONS = 15;
   unsigned long lastCheckMillis = 0;
   unsigned long sameSideCounter = 0;
   ```

   Just in case you are wondering why we are using `unsigned long` as a data type instead of `int` here, we will make use of the `millis` function, which returns the number of milliseconds since the Arduino started. When the Arduino has been running for a while, this number can get really big. `unsigned long` variables can store bigger (positive) numbers than `int`.

   The maximum number an `unsigned long` variable can hold is `4,294,967,295`. When used to store milliseconds, this equals roughly 49 days of using the Arduino continuously. You can read more about it in the Arduino reference (`https://www.arduino.cc/reference/en/language/variables/data-types/unsignedlong/`).

2. We also need to create another variable to hold the resting side ID. Once we have detected that the cube has been lying on the same side for a while, this variable will hold the side ID. Place this directly under the variables you just added:

```
int currentRestingSide = -1;
int previousSide = -1;
int currentSide = -1;
```

3. Next, we will implement the actual check. Under our if...else conditions (at the end of the loop function), remove the following line:

```
Serial.print("Side: "); Serial.println(currentSide);
```

4. Now add the following lines of code instead:

```
if (currentSide != currentRestingSide) {
  if (previousSide == currentSide) {
    unsigned long currentMillis = millis();
    if (currentMillis > lastCheckMillis + CHECK_MILLIS) {
      sameSideCounter += 1;
      lastCheckMillis = currentMillis;
    }
  } else {
    sameSideCounter = 0;
  }
  if (sameSideCounter >= CHECK_ITERATIONS) {
    // the side changed and it is in a stable state
    currentRestingSide = currentSide;
    Serial.print("Side changed: ");
Serial.println(currentRestingSide);
  }
}
previousSide = currentSide;
```

Let me explain what we are doing here. After we update the currentSide variable in our if-else checks, we compare it to currentRestingSide. This variable will hold the side ID of the side that passed our test to eliminate false detection when rotating the cube (the sides in between). We then check whether, in the last loop iteration, the side ID was the same as it is now. If it is, we store the current time (the time the microcontroller has been running for in milliseconds) in the currentMillis variable.

5. As we want to check every 50 milliseconds (stored in the CHECK_MILLIS variable) whether the side is still the same, we do this using the following line of code. The code inside this condition will be run if the last time we checked is longer than 50 milliseconds ago:

```
if (currentMillis > lastCheckMillis + CHECK_MILLIS) {
  sameSideCounter += 1;
  lastCheckMillis = currentMillis;
}
```

6. We then increment the sameSideCounter variable by 1. We need this variable to check whether the cube has been resting on the same side for a while. The last statement in the if condition updates the lastCheckMillis variable. We just performed the check, so we do not need to do it again in the next 50 milliseconds. The next time the loop function runs, it will not enter this if condition, because the loop function is executed much more often than every 50 milliseconds:

```
if (previousSide == currentSide) {
  ...
} else {
  sameSideCounter = 0;
}
```

We just had a look at what happens inside the outer if condition. The counterpart is easier. If the side ID of the previous iteration is not the same as the currently detected side, we just reset our side counter, sameSideCounter.

7. After we have either incremented or reset our counter, we perform the following check:

```
if (sameSideCounter >= CHECK_ITERATIONS) {
  // the side changed and it is in a stable state
  currentRestingSide = currentSide;
  Serial.print("Side changed: ");
  Serial.println(currentRestingSide);
}
```

If our counter is bigger than 15 (stored in CHECK_ITERATIONS), we detect a side switch. You could read our conditions as follows: if the currently detected side is different from the last resting side and the cube is lying on the same side for 750 milliseconds, the side was changed.

But where is 750 coming from?

CHECK_ITERATIONS is 15, and CHECK_MILLIS is 50. And 50 times 15 is 750. So, it takes 750 milliseconds to detect a side change.

 Increasing CHECK_ITERATIONS might increase the detection quality, but it will take longer to detect a side switch.

The code until here can be found in the book's repository under ch7_04.

Summary

In this chapter, we learned how to build an orientation sensor using tilt switches from scratch and saw how to improve it step by step. We discussed how to hide the complexity of our project by building a cube around it to make it presentable.

In the next chapter, we will build upon this by adding MQTT into the mix. The cube will be enhanced to broadcast the associated activity along with the stopped time, for example, **Watching TV: 123 minutes**.

We will also have a look at how to make use of MQTT clients for Android and iOS to display the data published by the cube.

Questions

1. How should tilt switches be arranged to give reliable readings?

Further reading

- If you want to know more about tilt switches, read this excellent article by Vidia Nindhita: http://blog.vidianindhita.com/2018/02/27/all-about-tilt-switches/.

8
Building a Smart Productivity Cube, Part 2

In the previous chapter, we built a smart productivity cube using a clever combination of mechanical tilt switches. The tilt switches make it possible to reliably detect which side the cube is standing on.

In this chapter, we will add MQTT into the mix and set up a third-party iOS and Android MQTT client to receive data from the cube. On your smartphone, you can then see, for example, **You have been watching TV for 123 minutes**.

Being able to use MQTT smartphone apps to send and receive MQTT messages opens up a lot of possibilities. With just a few clicks, you will be able to create your own app-dashboards to display and—in future projects—control your physical MQTT devices.

After completing this chapter, you will be able to create prototypes using MQTT that interact with smartphone MQTT clients such as **IoT OnOff** for iOS and **MQTT Dash** for Android. You will be able to create your own dashboards to send information to your device or visualize published text coming from your device.

To facilitate the learning process, this chapter is divided into the following sections:

- Making your device smart
- Displaying the activity on your smartphone

Making your device smart

The detection of all of our cube sides is working now. Let's add MQTT to our code base:

1. Open the MQTT example code from this book's repository: `general/arduino/mqtt_shiftr_send_receive_example`.

   As we did in the other projects, let's run this test-code first before we integrate it into our main sketch to make sure the Wi-Fi, as well as the MQTT connection, is working.

2. Enter your Wi-Fi name (`WIFI_SSID`) and Wi-Fi password (`WIFI_PASSWORD`). You can also customize `MQTT_DEVICE_NAME`. You will see it on shiftr.io when you look at the network graph.

3. Pick any name you like here, but stick to the characters a-z, 0-9, and dash (-). Other characters might work as well but might also lead to problems with certain MQTT clients or servers.

4. Try to make your device name unique (for example, by appending a few random numbers). The device name should not be longer than 23 characters for best compatibility:

   ```
   const char WIFI_SSID[] = "your network name here";
   const char WIFI_PASSWORD[] = "your network password here";
   const char MQTT_SERVER[] = "broker.shiftr.io";
   const int MQTT_SERVER_PORT = 1883;
   const char MQTT_USERNAME[] = "try";
   const char MQTT_PASSWORD[] = "try";
   const char MQTT_DEVICE_NAME[] = "hellomqtt"; // can be freely
   picked
   ```

Let's now refer to the following instructions and make our project smart:

1. Upload the code and you should see the following:

   ```
   .
   Connected to WiFi!
   Connecting to MQTT server...
   Connected to MQTT server
   incoming: /tims-channel - hello
   incoming: /tims-channel - hello
   incoming: /tims-channel - hello
   ...
   ```

   If you see incoming, ..., being printed, you know that the connection to the MQTT server shiftr.io is working. Great! Now let's integrate the bits we need.

From here on, we will extend the Arduino sketch that we started to work on in the previous chapter (Chapter 7, *Building a Smart Productivity Cube, Part 1*). If you had any problems following along, you can open the sketch from this book's repository in the `ch7/arduino/ch7_04` folder and use it as a starting point.

2. Copy and paste the following `include` statements and variable definitions into your main sketch (to the top) and make sure you do not overwrite your network credentials with the placeholder ones (`"your network name here"`), but use your actual username and password:

```
#include <WiFiNINA.h>
#include <MQTT.h>

const char WIFI_SSID[] = "your network name here";
const char WIFI_PASSWORD[] = "your network password here";
const char MQTT_SERVER[] = "broker.shiftr.io";
const int MQTT_SERVER_PORT = 1883;
const char MQTT_USERNAME[] = "try";
const char MQTT_PASSWORD[] = "try";
const char MQTT_DEVICE_NAME[] = "hellomqtt"; // can be freely
picked

int status = WL_IDLE_STATUS;

WiFiClient net;
MQTTClient client;
```

3. Next, copy the complete `connect` function over to your main sketch. Place it at the very end of your sketch. The first part takes care of establishing a Wi-Fi and MQTT connection:

```
void connect() {
  // first connect to the wifi
  Serial.print("Checking wifi...");
  while (status != WL_CONNECTED) {
    status = WiFi.begin(WIFI_SSID, WIFI_PASSWORD);
    Serial.print(".");
    delay(1000);
  }
  Serial.println(); Serial.print("Connected to WiFi!");
Serial.println();
  // second connect to the MQTT server
  client.begin(MQTT_SERVER, MQTT_SERVER_PORT, net);
  Serial.println("Connecting to MQTT server...");
  while (!client.connect(MQTT_DEVICE_NAME, MQTT_USERNAME,
```

```
MQTT_PASSWORD)) {
    Serial.print(".");
    delay(1000);
}
```

The second part of the `connect` function gets executed once the connection to the MQTT server has been established:

```
Serial.println("Connected to MQTT server");
// define what should happen when messages are incoming
client.onMessage(messageReceived);
// subscribe to MQTT topics
client.subscribe("/tims-channel");
}
```

4. Now, copy the `messageReceived` method over as well, to the very end of your sketch.

5. Next, we need to integrate the code from the `setup` function. If you compare the `setup` function of our main sketch and the `setup` function of the MQTT example sketch, you will see that we already have the following line in our main sketch:

```
Serial.begin(115200);
```

6. So, there is just one line of code left, which we need to copy to the beginning of the `setup` function in our main sketch—the `connect()` call. Place it at the end of the `setup` function:

```
void setup() {
  Serial.begin(115200);
  pinMode(SWITCH_PIN1, INPUT_PULLUP);
  pinMode(SWITCH_PIN2, INPUT_PULLUP);
  pinMode(SWITCH_PIN3, INPUT_PULLUP);
  pinMode(SWITCH_PIN4, INPUT_PULLUP);
  connect();
}
```

7. The only thing left is the `loop` function. We need the first few lines, which make sure that the connection to our Wi-Fi network, as well as the MQTT server, is active:

```
client.loop();
if (!net.connected()) {
  connect();
}
```

8. Place the code at the beginning of your `loop` function in the main sketch:

```
void loop() {
  client.loop();
  if (!net.connected()) {
    connect();
  }
  bool s1 = digitalRead(SWITCH_PIN1);
  bool s2 = digitalRead(SWITCH_PIN2);
  ...
```

Every time the cube is placed on a different side, we want to start a timer and publish the time via MQTT. We also want to assign an activity name to each side. You could, for example, use the bottom side to record the time you spend exercising, the top side to record how long you spend reading the news, and the other sides to record how long you work on projects.

MQTT messages should not be published too often, because you might get kicked out of the MQTT server. Every MQTT server defines its own limits. You can find the limits of shiftr.io here: `https://docs.shiftr.io/guides/limitations/`.

If you completed the first and second project in this book (and bought not only one but three Arduino MKR WiFi 1010), you could also make them all talk to each other.

With very minimal modifications, you could display the activity you are working on the e-paper display. To utilize the smart (pet) feeder, you could fill it with candy and, if you work *x* minutes on the `complete Arduino project` activity, release some candy. Yummy!

This is why I love MQTT. It is very composable. Combining various Arduino-based microcontrollers that send and receive MQTT messages is a piece of cake.

Let's add the logic needed to stop the time and publish it via MQTT every 10 seconds by following the instructions:

1. First, we need a variable for each side to hold the time. Create the following variables at the top of your sketch, right before the `setup` function:

```
unsigned long sideMillis[] = {0, 0, 0, 0, 0, 0};
unsigned long sideTimerStartMillis = 0;
String activityNames[] = {
  "Exercising",
  "Eating",
  "Watching TV",
  "Working on Arduino",
```

```
    "Social Media",
    "Sleeping"
};
```

The first array will be used as the time store for the recorded times. If we place the cube on a side on which it was lying before, it will not start at 0, but use the previous time and add up to it. For this, the `sideMillis` array is used—it holds one timer for each side.

The second variable, `sideTimerStartMillis`, will be used to hold the time in milliseconds since the cube was placed on it. Once the sides change, this will be used to calculate the time the current side was placed on and the resultant time will be added to one of the timers in the `sideMillis` array.

The third variable, `activityNames`, is an array of strings that holds the activity names we want to associate with any of the six sides. Feel free to customize it to your own needs. You just have to make sure that it contains exactly six string values and not less.

2. We now need to create the timer logic. Scroll down until you reach the code that gets executed when the side changed:

```
if (sameSideCounter >= CHECK_ITERATIONS) {
  // the side changed and it is in a stable state
  currentRestingSide = currentSide;
  Serial.print("Side changed: ");
Serial.println(currentRestingSide);
}
```

3. Following the `Serial.println` call, add the following lines:

```
if (previousSide != -1) {
  unsigned long activityMillis = currentMillis -
sideTimerStartMillis;
  sideMillis[previousSide] += activityMillis;
  Serial.print("Seconds spent on ");
  Serial.print(activityNames[previousSide]);
  Serial.print(": ");
  Serial.print(sideMillis[previousSide] / 1000); // seconds
  Serial.println(" seconds");
}
sideTimerStartMillis = currentMillis;
```

4. We divide the milliseconds stored in `sideMillis[previousSide]` by `1000` to get seconds. If you prefer minutes, you can just divide the resultant number by 60:

```
Serial.print(sideMillis[previousSide] / 1000 / 60); // minutes
```

Because we need to change sides regularly while working on the code, using seconds is more practical. But later on, feel free to change it to minutes.

5. If you compile the code now, you will get an error message:

```
'currentMillis' was not declared in this scope
```

6. We need to access the current time (in milliseconds) in this part of the code. We already created a variable to hold the current milliseconds in the preceding few lines:

```
unsigned long currentMillis = millis();
```

But as it is in a different scope (inside an `if` clause), we cannot use it directly in the code we are working on right now.

7. Let's move it to the beginning of our `loop` function so that it becomes visible in our scope. After you move the `currentMillis` definition up, the beginning of the `loop` function should look like this:

```
void loop() {
  unsigned long currentMillis = millis();
  client.loop();
  ...
```

`currentMillis` can now be used everywhere in the `loop` function.

8. Recompile the code. It should compile without problems now. Upload it to the board.

If you look at the serial monitor and rotate the cube, you should get an output like this:

```
Time spent on Watching TV: 1 seconds
Side changed: 4
Time spent on Social Media: 3 seconds
Side changed: 0
Time spent on Exercising: 9 seconds
Side changed: 4
Time spent on Social Media: 7 seconds
```

```
Side changed: 0
Time spent on Exercising: 11 seconds
...
```

9. Now let's publish this information via MQTT.

    There are multiple places in our code where it would make sense to publish the
    message. We could, for example, publish a message every time we detected a side
    change (after it passed the check that it is in a stable state). But if we want to
    display the time we are spending on the current activity (for example, on the e-
    paper display that we built in project 2 or an MQTT smartphone client), we do not
    get enough updates. We would just receive an MQTT message once the side
    changed but would not have access to the updated time.

    We need to create another timer that publishes the current task and time
    cyclically. Every 10 seconds is a reasonable time for that.

10. At the top of your sketch, right before the `setup` function, create two new
    variables, `PUBLISH_MILLIS` and `lastPublishMillis`:

    ```
    int PUBLISH_MILLIS = 1000 * 10; // 10 seconds
    unsigned long lastPublishMillis = 0;

    void setup() {
      ...
    ```

11. Go to the end of your `loop` function and add the following code
    (before `previousSide = currentSide;`):

    ```
        ...
      if (currentRestingSide != -1) {
        if (currentMillis > lastPublishMillis + PUBLISH_MILLIS) {
          unsigned long activityMillis = currentMillis -
    sideTimerStartMillis;
          unsigned long timeSpend = (sideMillis[currentRestingSide]
    + activityMillis) / 1000;
          String text = activityNames[currentRestingSide] + ": " +
    timeSpend + " seconds";
          Serial.println(text);
          client.publish("/tims-channel/cube/activity", text);
          lastPublishMillis = currentMillis;
        }
      }
      previousSide = currentSide;
    }
    ```

12. Reupload the code and you should see an output every 10 seconds with the time spent on the specific activity in the serial monitor.

    The code until here can be found in this book's repository as `ch8_01_mqtt`.

    Now, let's find out whether the detected side is being successfully published to the MQTT server.

13. Visit the following URL in your browser: `https://shiftr.io/shiftr-io/try/topic/tims-channel/cube/activity`.

If you used a different topic name, you have to change the URL accordingly.

You can also get there manually by navigating to `https://shiftr.io`, clicking on **Try** in the website's menu, and then clicking on the node activity, which is connected to the `tims-channel` node (feel free to use another channel; you need to replace all occurrences of `tims-channel` in your code then).

The public namespace, which you publish and subscribe to, is shared by all users on shiftr.io. It might be hard to find the specific topic/nodes in there because it is so crowded. Also, other readers of this book might use the same topic name.

Using a private account and namespace later on is highly recommended:

shiftr.io   Try   Explore   Get Started   Documentation

shiftr-io/try/topic/tims-channel/cube/activity

| Messages | Number Series | Publish Rate |

Exercising: 20 seconds
tims-channel/cube/activity - hellomqtt - 22 Bytes - May 09, 2019 02:00 - Q0 NR

Exercising: 10 seconds
tims-channel/cube/activity - hellomqtt - 22 Bytes - May 09, 2019 02:00 - Q0 NR

The topic view on shiftr.io reveals all recent MQTT messages published to the specific topic

While the shiftr.io website is great for topic inspection and viewing published messages in a log-like view, we might want to check on our phones how long the activity is taking. Luckily, there are many options available. The one with the least effort necessary is certainly one of the many iOS and Android MQTT clients. Let's try them out to inspect the data we are publishing.

In the previous chapters, we mostly made use of the command-line MQTT client, Mosquitto. In the next section, we will try out two popular MQTT clients for iOS and Android. But it is always good to have different ways to subscribe to the data. Do you still remember how to do it on the command line with Mosquitto? If not, have a look at the previous chapters. The command to use is `mosquitto_sub` to subscribe to a topic.

Displaying the activity on your smartphone

One of the great things about MQTT is the number of available clients that we can choose from. There are clients for Windows, Linux, macOS, Android, iOS, Arduino, and the web—basically everything.

In the following section, let's have a look at two popular MQTT clients for iOS and Android: MQTT Dash and IoT OnOff.

Using MQTT on Android via MQTT Dash

If you are not using an Android smartphone, feel free to skip this section.

MQTT Dash is an MQTT client for Android that is all about creating dashboards. You can have various dashboards that display incoming information (of topics you subscribed to) and where you can choose from a range of user interface widgets, which make publishing data very easy:

1. First, we need to install MQTT Dash. You can find it in the Android Play Store using its name, `MQTT Dash`, or using this link: `https://play.google.com/store/apps/details?id=net.routix.mqttdash`.
2. Once the installation has finished, run the app.

Setting up MQTT Dash and subscribing to the `/tims-channel/cube/activity` topic is very straightforward.

The following are detailed instructions on how to set it up to work with our smart productivity cube. Depending on when you read this book, the interface might have changed a bit:

1. To display the current activity of the smart productivity cube, we first need to create a new dashboard. Click on the plus icon in the top-right corner:

The home screen of MQTT Dash showing all your dashboards (yours will be empty because you do not have any)

2. In the next screen, we need to provide the following information, which is the minimum needed for any password-protected MQTT server. The shiftr.io namespace with the `try` username and `try` password is shared by all users, but it still uses a username and password, so we need to specify it:

   - **Address:** `broker.shiftr.io`.
   - **Port:** `1883` (default for MQTT).
   - **Username:** `try`.
   - **Password:** `try`.
   - **Client ID:** This is the name of your device. It needs to be unique on the server, for example, `your-name-648568`.

Adding a few random numbers at the end of your client ID decreases the chances that someone else picked the same name.

3. Now, let's input the missing fields into MQTT Dash to create a new dashboard.

The first checkbox can be ticked if you only want to use one dashboard for all your projects on shiftr.io. In my case, I leave it unchecked because I use multiple servers and prefer to manually connect to them:

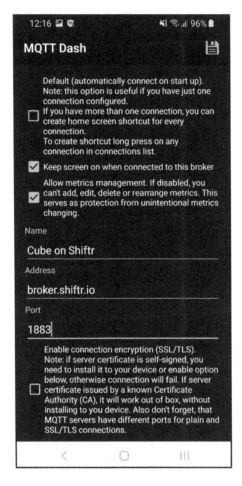

New dashboard screen

The second checkbox, **Keep screen on when connected to this broker**, makes sure that your screen will not automatically shut off. Leave it checked.

The third checkbox, **Allow metrics management**, can be left checked.

4. Next, you need to specify a name for your dashboard. I named mine `Cube on Shiftr`. You can identify your dashboards by their names on the home screen of MQTT Dash. This name is only used by the app and not sent to the MQTT server.

5. To successfully connect to the MQTT server, you need to tell MQTT dash which server it should connect to; in this case, the address is `broker.shiftr.io` and the port is the same as the default one, `1883`. You can find this information if you look at the shiftr.io MQTT reference (`https://docs.shiftr.io/interfaces/mqtt/`). Every other public MQTT server will have documentation with this information as well, in case you want to use another MQTT server at some point.

6. The next field, **Enable connection encryption (SSL/TSL)**, can be left unchecked. In a serious application, we definitely would want to secure our communication as much as possible, but in our case, it is all about prototyping and choosing the easiest path.

7. If you scroll down a bit, there are a few more fields to fill out:

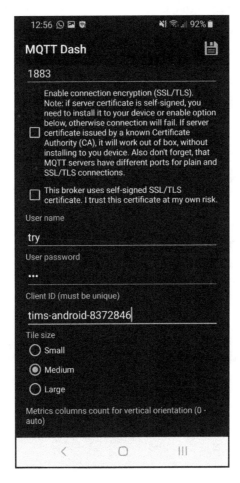

More settings for the new dashboard

8. Fill in `try` for the username and `try` for the password. If, later on, you want to use a private namespace, this is where you would enter your personal username and password for shiftr.io.

9. Last but not least, we need to fill in the **Client ID**. This is what we will see on the shiftr.io interface when we publish information to the MQTT server. This name is public and needs to be unique on the MQTT server. Restrict yourself to lowercase characters (a-z), dashes (-), and numbers for maximum compatibility with other MQTT clients. Also, it should not be longer than 23 characters.

   The following settings (tile size and metrics) are only important for the arrangement of the interface widgets on the dashboard and can be left as they are.

10. We are all set. Click on the save icon in the top-right corner and you should find yourself on the app's home screen, the dashboard overview:

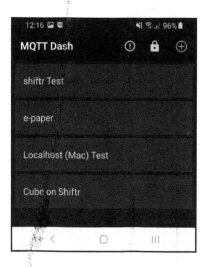

In the dashboard overview is a new entry Cube on shiftr.io

If you ever want to change the settings provided earlier, you need to long-press on the dashboard entry in this screen (the dashboard overview) and choose **Edit**.

11. Now, click on your newly created dashboard to open it:

Your newly created dashboard is still empty because it does not contain any widgets

12. Currently, you will just see an empty screen. We first need to create some interface widgets to display information. Click on the plus icon in the top-right corner to add a widget:

Available widget types in MQTT Dash

You have the following options available:

- **Text**: Display incoming text or publish text from your phone.
- **Switch/button**: Toggle between two states, like a light switch.
- **Range/progress**: Send a value in a certain range, for example, to set the brightness of an LED strip.
- **Multiple choice**: This is an easy way to send predefined strings; you select one of the various options to send.

- **Image**: This displays an image that you could send as a URL string, for example, the Arduino logo (`https://upload.wikimedia.org/wikipedia/commons/8/87/Arduino\_Logo.svg`). You could, for example, use it to show a workout image on the display of your phone when the recorded activity of the cube is `workout` or show an image of a TV when the recorded activity is `watch TV`.
- **Color**: Control the color of an LED strip.

Because we just want to display text, for now, choose **Text**.

13. On the next screen, we need to enter the name of the widget and the channel where the text we want to display is published to:

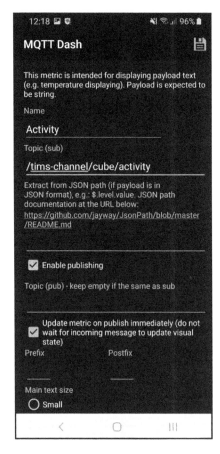

Text widget settings

14. Fill in `Activity` in the **Name** field and `/tims-channel/cube/` activity (or the topic you are using) in the **Topic (sub)** field.

> **Sub** here stands for **subscription**.

15. Scroll down a bit and set **Main text size** to **Small**. The default setting, **Large**, might lead to the text being cropped:

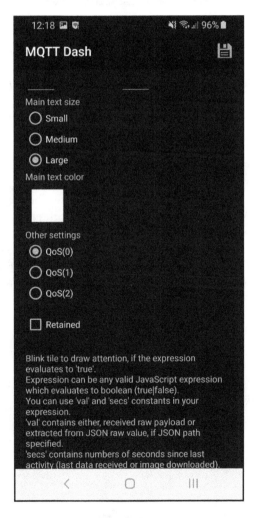

Text settings

16. We can leave all other settings as they are. Click the save icon in the top-right corner and you are brought back to the widget overview of our dashboard:

Our newly created text widget displays the current activity published by the cube

In the next section, we will get to know IoT OnOff, another powerful MQTT app.

Using MQTT on iOS with IoT OnOff

In this section, we will have a look at the app IoT OnOff (`https://www.iot-onoff.com/`). It is an Android as well as an iOS app. I will walk you through the process of using it together with the smart productivity cube from an iOS perspective. There are various other apps around, both on iOS and Android, but IoT OnOff offered the most interface widgets. Sadly, not all the features are free, but the free version is enough for what we are about to do with it. On the app's website (`https://www.iot-onoff.com/`), you will find a link for both versions of the app—the iOS version in the App Store, as well as the Android version in the Play Store. Install the app for your platform and accept the disclaimer (after reading it, of course). Then, follow these steps:

1. Enter a client ID. The client ID will identify your phone using IoT OnOff on the MQTT servers. You can freely choose the name, but there are some restrictions. For maximum compatibility to other MQTT servers and clients, your name should only use regular characters (a-z and 0-9) and hyphens (-). It also should not be longer than 23 characters.

A good name, for example, is `tims-phone-83746289`. The last part is just a random sequence of numbers, which will decrease the chances that your client ID was picked by another user as well. You could also use different word combinations that appear unique to you—for example, `tims-phone-iotonoff` is rather unlikely to be taken:

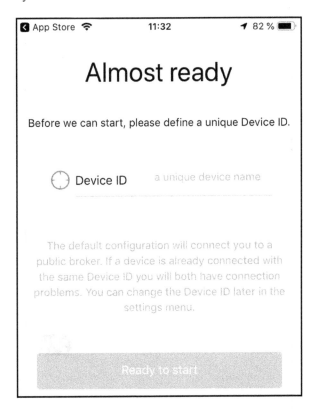

Device ID form

Once you enter a client ID, you will be forwarded to the next screen, where you can see an example dashboard to give you an idea of what is possible with the app. Similar to MQTT Dash, which we used in the previous section, IoT OnOff has a predefined set of interface widgets, which can either display incoming MQTT messages or be used to publish them:

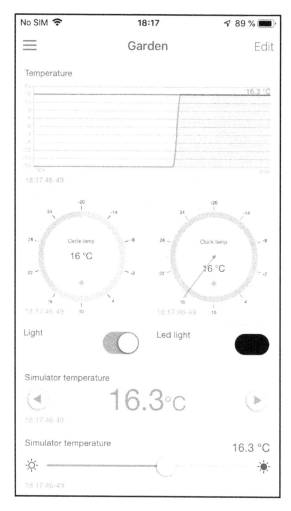

Dashboard overview in IoT OnOff

2. Now, click on the burger icon (the three lines) in the top-left corner to open the menu. Here, you can see all of your dashboards. Currently, there are only demo dashboards, which were predefined by the app developers.

3. Click on Edit in the top-left corner. On the dashboard overview, click on Add in the bottom-left corner to add a new dashboard:

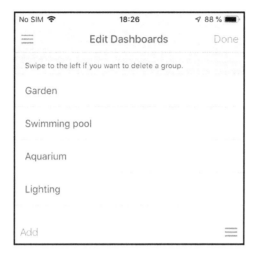

Dashboard edit overview

4. Provide a name for your new dashboard (I am using `Cube`):

Dashboard name input

The workspace name is not really important here and will only be used inside the app, not on the MQTT server.

5. Press **Ready** in the top-right corner. Afterward, you are back at the **Edit Dashboards** overview, where you will find a new entry, **Cube** (or whichever name you picked).
6. Click on **Done** to leave the dashboard edit mode:

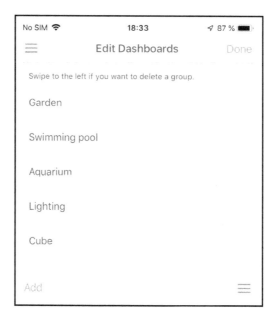

Edit Dashboards overview with new entry, Cube

7. Now click on the burger icon again.

In case you are wondering, the icon with the three lines is really called the burger icon because it looks a bit like a hamburger. Every user interface designer will understand what you are talking about when you mention the burger icon.

8. In the side drawer, click your newly created dashboard, **Cube**, now to open the dashboard:

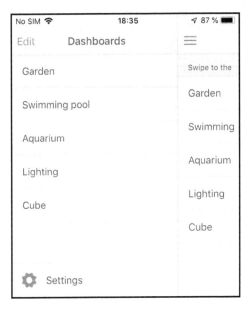

Side drawer revealing the newly created Cube dashboard

You will be presented with a white screen. This is because there are no interface widgets yet. We first have to create one:

Your Cube dashboard does not contain any widgets yet

9. Click on **Edit** in the top-right corner to activate the edit mode:

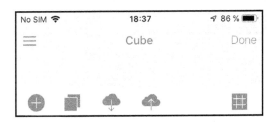

Dashboard edit mode

10. Now click on the + icon at the bottom to add a new interface widget. You will see an overview of all available widgets. Some of them might require you to buy the pro version of the app. We will stick to the free ones.

11. Click the message widget. We will use this widget to output the text sent from our Arduino:

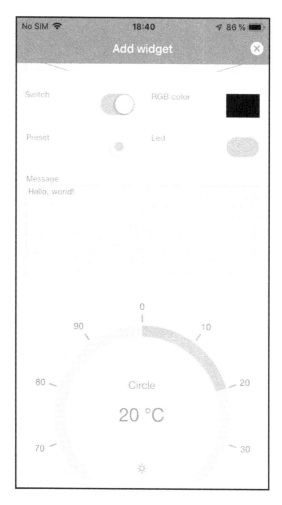

Available widgets in IoT OnOff

In the following screen, we need to provide the necessary information, so our widget can display the published message coming from the smart productivity cube.

12. Pick a name for your widget, for example, `Cube activity`:

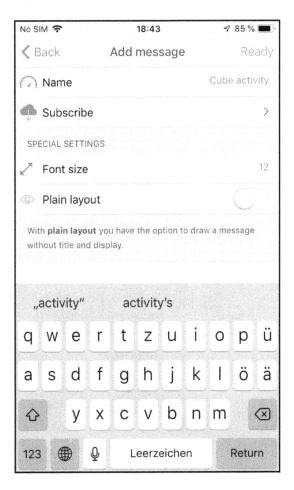

Widget settings

13. Now, click on **Subscribe**. Here, we will specify which topic to subscribe to. Enter the same topic name you used to publish information to and from the cube:

```
/tims-channel/cube/activity
```

14. Click the **X** icon in the top-right corner when you are done. Here, the app is very unintuitive. Clicking an **X** icon in such a situation symbolizes a cancel action. But not in this case. Clicking it will not discard the topic string you just entered but save it:

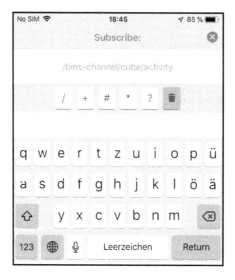

Widget subscribe settings

15. You will find yourself back at the **Subscribe** view.
16. Click **Ready** to save it.
17. Click **Ready** in the top-right corner once more to confirm your changes.
18. Your newly created widget was created and placed on your dashboard.

19. Press **Done** to leave the edit mode:

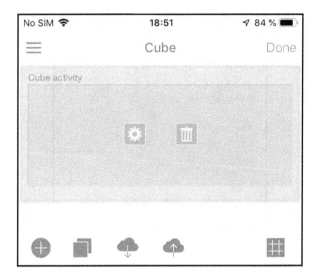

Dashboard edit mode with the newly created text widget

You might see a red warning message now saying that the **Free version is limited to 10 widgets**, as shown in the following screenshot:

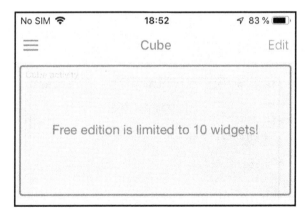

Limitation of the free version of IoT OnOff

We, therefore, need to delete the demo widgets before we can continue.

20. Click on the burger icon in the top-left corner to access the menu:

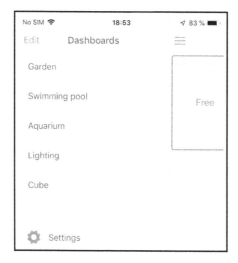

Drawer revealing your dashboards and the settings button

21. Click on **Edit** in the top-left corner.
22. To delete the demo dashboards (all except our **Cube** dashboard), swipe the dashboard entries to the left until they reveal a **Delete** button.
23. Press the **Delete** button.
24. Repeat swipe and delete for all of the demo dashboards so that we end up with only our **Cube** dashboard:

Edit dashboards overview—everything except Cube has been deleted

25. Click **Done** in the top-right corner.
26. Click the burger icon in the top-left corner to go back to the menu drawer:

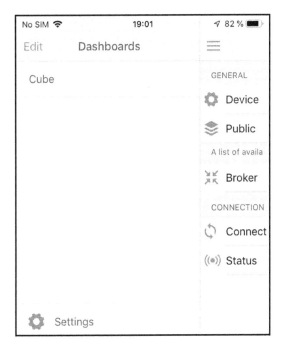

Side drawer revealing our dashboards and the settings button

We are nearly there. The next time you create a new widget in IoT OnOff, it will be easier—I promise!

You might be wondering how the app should know which MQTT server and what port, username, and password to use. It does not. This is the last step before we can display the information coming from the cube.

27. In the side menu, press **Settings** on the bottom.
28. In the settings menu, click **Configuration**.

29. Click on **Broker**:

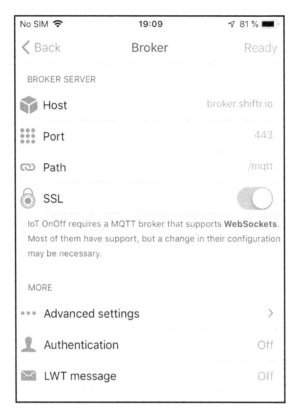

MQTT server settings

MQTT broker and MQTT server are the same. The term MQTT server is officially the correct one.

This looks familiar, right?

30. Enter `broker.shiftr.io` as host and `443` as port. Normally, we would use `1883` as the port, but IoT OnOff is using WebSockets to communicate with the MQTT server, therefore we need to specify another port number. You can also find out about the correct port to use by looking at the shiftr.io interface documentation (`https://docs.shiftr.io/interfaces/mqtt/`). In this case, we are using secure WebSockets, which requires that the **SSL** toggle is checked.

31. Scroll down a bit, then click on **Authentication**.
32. Enter `try` as username and `try` as password. Also, enable the toggle **Use authentication**:

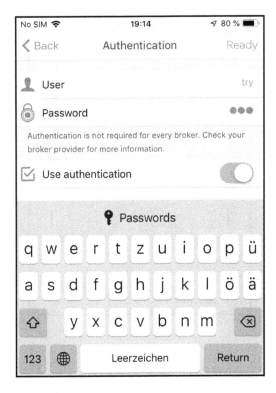

Authentication settings

We are ready to go now.

33. Click **Ready**, which will bring you back to the main **Settings** screen:

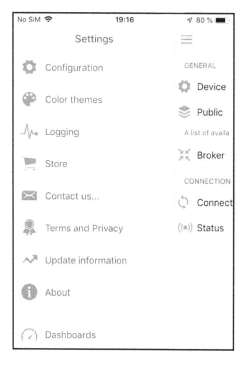

Main settings revealed inside the drawer

34. Click the burger icon to open the drawer, then click **Dashboards**.
35. Click the name of your dashboard, in my case, **Cube**:

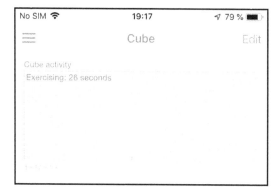

The text widget on our Cube dashboard shows the current activity, published by the cube

Finally! After a few seconds, you should see text coming in from your cube.

I know it took a while to set everything up, but adding new widgets is easy now. We don't have to set up our MQTT server, port, username, and password again.

You only need to specify which kind of widget you want to add and which topic the widgets should subscribe to receive the information.

Summary

In this chapter, we enhanced the code of our smart productivity cube, added MQTT to the mix, and learned how to make use of two popular MQTT smartphone apps.

First, we added a stopwatch functionality to our cube. Each side of the cube has its own timer and can be used as an individual stopwatch.

We then learned how to integrate the Arduino MQTT library into our code and how to periodically publish the current side the cube is resting on, as well as the resting time, for example **Watching TV: 123 minutes**.

Last but not least, we learned how to receive this information on our smartphones, using the third-party apps MQTT Dash and IoT OnOff. Once set up, they can be used to interact with your MQTT prototypes very easily.

I hope you don't stop here, but feel inspired by the vast variety of options and how you can create your own dashboards to display MQTT messages and publish MQTT messages via your smartphone.

Take your time to explore the apps and improve the prototype. In `Chapter 9`, *Presenting Your Own Prototype*, we will discuss at length how to present your prototype so that it looks good.

Questions

1. Can the MQTT clients for Android and iOS just be used to display MQTT messages?
2. How can you send just the activity time, without the activity name?
3. What makes a good client ID?

Further reading

- Feel free to try the web-based MQTT client by HiveMQ as well: http://www.hivemq.com/demos/websocket-client/

3

Section 3: Prototypes

This section discusses different ways to make prototypes look more professional, including building cases using technologies such as laser cutting and 3D printing.

The following chapter will be covered in this section:

- `Chapter 9`, *Presenting Your Own Prototype*

Presenting Your Own Prototype

9

In the previous chapters, you learned how to build three prototypes: a smart pet food dispenser, a smart e-ink to-do list, and a smart productivity cube. These chapters mostly explained the technical side—how to make it work. We did not care too much about how the prototype looked, but this is important as well. Especially if you want to present your prototype to other people.

In this chapter, we will have a look at different techniques that can be used to create a case for your prototype. We will also have a glimpse at batch production and how to take your prototypes to the next level.

The following topics will be covered in this chapter:

- Using household boxes as cases
- Using laser cutting to build custom cases
- Using 3D printing to build custom cases
- Evolution of a circuit – from breadboard to PCB

Using household boxes as cases

Building a good-looking case does not need to be hard. In the hands-on chapters of this book (Chapter 5, *Building Your Own Automatic Pet Food Dispenser*, Chapter 6, *Building a Smart E-Ink To-Do List*, Chapter 7, *Building a Smart Productivity Cube, Part 1*, and Chapter 8, *Building a Smart Productivity Cube, Part 2*), we built all of the cases from scratch using cardboard. While this gives you a lot of flexibility, it might take more time than you are willing to spend for your next Arduino-based project.

A great starting point for custom-built prototype cases are household boxes or boxes that you can buy in arts and crafts stores. Most prototypes live in a rectangular box anyway. Why not use an existing one and put in the holes needed for your user interface, cables, or display? Using existing boxes also lets you re-use boxes that might go into the trash otherwise. It would be good to find a second use case for those.

There are many things in a typical household to be considered for prototyping cases. Scan your environment for boxes that meet the following criteria:

- **Availability**: Use whatever boxes you can easily get. Maybe you have a space where you keep the packaging of electronics that you bought? This is the first place where you should look for a candidate for a case.
- **Stability**: Depending on your use case, it must be stable. In many cases, cardboard is enough for a first prototype, but plastic gives you more stability. For maximum stability, aluminum is recommended.
- **Workability**: Find a material that you can work with using the tools at your disposal. A drill is your best friend when working with materials harder than cardboard.
- **Closability**: It is very handy if a box storing a prototype can be opened and closed. In case you have a loose cable or you want to upgrade your prototype, you have easy access to the hardware. But having a visible lid might not look best, so here you have to decide between looks and function.
- **Size**: The case must be big enough to house the breadboard, as well as the connected electronics.
- **Conductivity**: Some materials are conductive or flammable. When using a box made out of anything but plastic, make sure to isolate all electric parts of your prototype accordingly, for example, using electrical tape. The pins of the Arduino or connected electric modules should not touch the case walls.

In the following, you can see a selection found in an arts and crafts store that can be used as prototype cases:

Quick prototypes using boxes / a picture frame from an arts and crafts store

The prototype on the left is made out of a picture frame. It took about 5 minutes to measure how big the holes should be, calculate the margins so everything is centered, cut the holes out using scissors, and, finally, attach the buttons. Looks great for a 5-minute prototype, doesn't it?

The original purpose of the hollow wooden case on the right is a pencil holder. I just turned it upside down and drilled a hole in its base. It also did not take more than a few minutes to convert this pencil holder into a nice-looking prototype case.

In the back, you can see some more boxes that work just great as prototype cases. Just cut out the holes to place your interface elements in and you are ready to go.

Designing the front-plate interface

Most physical prototypes have some sort of user-facing interface, typically on the top of the device. If your prototype only has one knob or button to control the inner workings, a text label explaining its function might not be needed. But what if you have three knobs? What if there are 20 user-facing control elements, knobs, buttons, and sliders? In that case, you absolutely need labels to explain what control element does what.

Personally, you might find it relatively easy to remember what the individual controls are for, but your friends and family will have a hard time controlling your prototype.

There is always the possibility of using pen and paper to explain the different control elements, but writing the labels by hand might not make the best impression. We want to produce good-looking prototypes here, so there must be a better way.

Following, you can find two possibilities for labeling your control elements:

- **Alphabet stickers**: Typically to be found in arts and crafts stores or big online stores, you can get the whole alphabet as stickers. Let's say you have a button that is used to "Reset" your prototype. In this case, putting the sticker with the character "R" (for "Reset") might do the trick to guide the users of your prototype. Next to alphabet stickers, you can also get so-called "Transfer Alphabet Stickers". These can be transferred to your prototype case, mostly using heat (for example, using an iron).
- **Printed interface graphics**: Using a 2D design tool, you can design a good-looking interface including labels for all of your interface elements. Using a printer, you can then print it out, glue it on top of your prototype case, and finally, drill or cut the holes needed for buttons, knobs, or whatever interface elements you are using. The following image is an example of a printed interface graphic to be placed in the frame shown in the previous image. The knob graphics are placeholders for the physical knobs. Using text labels makes it clear what each control element is doing:

Example of a printed interface graphic

There is also a trick combining both techniques: using a 2D graphics application, you can design a black and white design for your interface, print it out using a laser printer (an inkjet printer will not work here), and then transfer it to your prototype. In the YouTube video, *5 Ways to Print on Wood | DIY Image Transfer* (https://www.youtube.com/watch?v=xHOWUR8vTvo), you will find various ways how to transfer images to a piece of wood; some of these methods will also work to transfer your design onto other materials.

Warning: Some of the methods for transferring prints as described in the previously mentioned YouTube video require the use of chemicals. Please make sure to inform yourself about the method being used before doing so and take the necessary precautions, for example, by wearing gloves and making sure there is enough ventilation.

To design the front-plate, you have many choices. Design applications must meet the following criteria to produce graphics to be used as physical interface explanations:

- **Work in cm/mm/in**: On computer screens, you typically work in pixels, which describe physical points on your screen. In the real world, we don't deal with pixels but cm and mm or inch and feet. The design application must support one of these units, so you can design the front-plate graphics in the same size as your physical prototype case. When printing a graphic for this use case, you also have to make sure to select **actual size** in the printing dialog, so as not to print in another scale (zoomed in or out), or your print will not fit on your physical case.
- **Basic shapes**: The design applications should support basic shapes such as rectangles and circles to give you some freedom while designing. Using rectangles around multiple interface elements can guide the user by creating visual interface element groups.
- **Text rendering**: The design application should be able to display text and optimally to rotate it. Rotating the text 45 or 90 degrees might make it easier to fit a lot of user interface elements next to each other.

Some professional vector applications for designing user interfaces for printing are Sketch, Adobe Illustrator, Affinity Designer, and Figma. If you do not own any of these applications, I recommend you try out Figma (`https://www.figma.com`), a modern web-based 2D design tool that can be used for free for up to three projects (this might change by the time you read this book).

When designing user interfaces, make sure to use the same distances between elements. It will look messy if you use different distances for the same kind of element. Also, creating groups of control elements might bring order into your design if you have a lot of interface elements.

I hope this section gave you some inspiration on how to quickly build the next case for your prototype. Keeping an eye out for good looking boxes is already 50% of what you need—but not only regular boxes: Tupperware or other enclosing objects might be of interest. If you have an arts and crafts store near your home or access to an online store in that field, I would highly recommend having a look at what boxes they have to offer.

If you want to get a little crazy, use non-rectangular objects. Recycle an old shoe as the home for your prototype, use an old planter, or a mug. There's nothing wrong with that. The weirder your prototype looks, the more memorable it will be for people seeing it.

Using laser cutting to build custom cases

Now that you know how to build a case using materials easily available at home, you may wonder whether other options can help you to make more professional-looking cases for your prototype. One option that can help you to achieve a nicer, sleeker look for your prototype case is laser cutting:

Laser-cut boxes made out of MDF sheets: the perfect base for electronic prototype cases (Stacked cubes by Jared Tarbell (https://www.flickr.com/photos/generated/) is licensed under CC-BY 2.0 (https://creativecommons.org/licenses/by/2.0/))

Laser cutting is a technology that uses a high-powered laser beam to cut different materials into custom shapes and designs. It works by generating a laser beam, focusing it onto the surface of the workpiece, and striking the surface with the beam to heat and melt it. In the process, flowing with the focused beam, there is usually an assist gas (that means, a cutting gas) that is used to cool the focusing lens and can be used to blow melted material out. Due to the precision and speed with which laser cutting allows users to cut a wide variety of materials (such as plastic, metal, wood, glass, and paper) with different levels of complexity, it is a very useful tool for building great cases for your prototype. The following is an example of a laser cut prototype case:

If you are interested in producing a case for your prototype using this technology, there are three main ways in which you can do it: buying a laser cutter, using one at a **fabrication laboratory** (**FabLab**), or using services that cut the pieces for you.

Buying a laser cutter

If you find yourself frequently building prototypes and you need the cases to be of high quality, buying a laser cutter is an option that you can consider. However, unlike 3D printers, there is little value in buying a laser cutter yourself:

- Laser cutters that cost a few thousand euros aren't close to the performance of industrial laser cutters that you can find, for example, in FabLabs.
- More affordable desktop laser cutters in the range of a few hundred euros are often hard to set up and not secure. Without a proper ventilation system, you will be exposed to gases that are produced during the cutting process.
- Some of the laser cutters in the cheapest price range are also not properly protected against misuse, possibly exposing your eyes to the dangerous laser beam.
- Many laser cutters directed at hobbyists have just enough power for engraving but cannot cut material that is thick enough to build a box for physical prototyping.

If you are serious about buying a laser cutter, there is much for you to learn and models to compare. A good starting point for available desktop laser cutters can be found on the website of miniFABLAB (`https://www.minifablab.nl/lasercutter/`).

FabLab

If you are like most of us and you do not have a lot of money to spend on a high-quality laser cutter, then a good option is accessing one through a **FabLab**:

Typical FabLab allowing members to use 3D printers and woodworking tools (FabLab VERITAS—Centro de Investigación para la Innovación, Universidad VERITAS (https://commons.wikimedia.org/wiki/File:Fab_Lab_VERITAS_-_Centro_de_Investigaci%C3%B3n_para_la_Innovaci%C3%B3n,_Universidad_VERITAS.jpg) by Rogarita is licensed under CC BY-SA 4.0 (https://creativecommons.org/licenses/by-sa/4.0/deed.en))

FabLabs are workspaces that are typically equipped with a variety of computer-controlled production technology tools such as 3D printers, laser cutters, CNC routers, and design software. In general, the purpose of these spaces is to enable people to make different things by granting them access to different equipment and sometimes by organizing events where people can learn how to use the different tools.

In general, FabLabs allow you to access expensive, good-quality equipment that you would most likely not be able to afford yourself, for a membership fee. Membership fees may vary depending on what equipment you are given access to. Cheaper membership options may restrict access to some of the equipment, so you want to inform yourself well to make sure that your membership type grants you access to the equipment that you would like to use.

If you are thinking about checking a FabLab, the easiest way to see whether there is a FabLab near you is to search for it online. However, you can also check this website, which offers a compilation of FabLabs around the world: `https://www.fablabs.io/labs/map`.

If you do not get any results searching for a FabLab in your home town, try searching for "maker space" instead.

Laser cutting services

Another option for making cases for your prototype with laser cutting technology is using a laser cutting service. Laser cutting services will basically cut the pieces in the way and material that you specify and ship them to you. You just need to upload your design, select the desired material, check the pricing, and place the order. It is a good option if you do not know how to use a laser cutter, you do not have access to one, or you just simply don't want to cut the pieces yourself.

Pricing varies greatly depending on your design (including the sizes of the pieces) and the material and its thickness. Usually, prices for cutting materials such as cardboard are the cheapest (for example, 2 EUR for an A4-sized wooden cardboard piece) and metal or wood the priciest (42 EUR for 244 cm x 122 cm and 6.5 mm thickness plywood piece). To get a detailed idea of how these services work and the prices, you can check some of the services listed following. Most of them are based in Europe or North America, so you may also want to search for a service in your country to have a better idea about prices there:

- **Cotter**: `https://cotter.co/material/aluminium/`
- **Ponoko**: `https://www.ponoko.com/`
- **Formulor**: `https://www.formulor.de/`
- **Sculpteo**: `https://www.sculpteo.com/en/lasercutting/`

The downside of using a web-service for laser cutting is the waiting time until the cut parts are delivered to your home. After you receive the parts, if your design has some issues, you have to modify your design accordingly, re-upload it to the laser cutting web-service, and, again, wait for it to be delivered.

How to generate a design for a laser cutter

Now that you have a good idea of what options are available for laser cutting pieces for your prototype's custom case, let's discuss how to generate the case's design for the laser cutter.

Since a laser cutter is a computer-controlled machine, you need to use 2D vector software to create the design of your pieces. To be able to use the 2D vector design software, however, you will need to learn how to use it. Due to the scope of this book, it is not possible to explain in detail how to use design software but following are some commonly used software programs and links with useful information on how to get you started if you are not familiar with them.

CorelDRAW

CorelDRAW (https://www.coreldraw.com) is a comprehensive graphic design software that allows you to create vectors, page layouts, and edit photos, among many other things. It produces high-quality designs but it is expensive. You need to either buy the program for about 700 EUR or get a subscription with an annual fee of approximately 240 EUR. For students, there is a reduced version that costs about 100 EUR. If you are interested in working with this software, here are two useful links that can help you to get a rough idea of how the program works and how to use it for designing simple prototype cases:

- **Beginner tutorial**: In case you are interested in using the program but do not have previous experience with it, the short beginner tutorial, *CorelDRAW X8 - Full Tutorial for Beginners*, on YouTube can help you to get a rough idea of how it works: https://www.youtube.com/watch?v=JygEl1AFN-M.
- **Designing boxes for laser cutting**: For a quick and easy tutorial on how to design a box for laser cutting, you can check the video, *Use CorelDraw Software to Draw a Box for Laser Cutting*, on YouTube: https://www.youtube.com/watch?v=2ijEqmfHGV4.

Adobe Illustrator

Adobe Illustrator (`https://www.adobe.com/products/illustrator.html`) is a powerful vector graphics software that lets you create logos, icons, drawings, typography, and illustrations for print, web, video, and mobile. This software is widely used for creating high-quality designs. The downside of this software is that, after the trial period, you will need to pay a subscription fee of approximately 20 EUR per month to be able to use it. If you have no previous experience with this software, here are some links that can help you to get started:

- **Illustrator tutorials**: Adobe offers a good selection of tutorials for both beginners and advanced users. The whole collection can be found at `https://helpx.adobe.com/illustrator/tutorials.html`.
- **The basics of Adobe Illustrator**: If you have never used Illustrator before, among the tutorials in the Adobe Illustrator page, this one might help you to understand the basics of it: `https://helpx.adobe.com/illustrator/how-to/ai-basics-fundamentals.html`.
- **Laser cutting with Adobe Illustrator**: For an easy tutorial on how to set up a design on Adobe Illustrator for laser cutting, you can check the tutorial, *Laser cutting with Adobe Illustrator*, on YouTube: `https://www.youtube.com/watch?v=FFK3VI7i6Eg`.

AutoCAD

AutoCAD (`http://www.autodesk.com/AutoCAD`) is a computer-aided design software that is used primarily by engineers and architects to create precise, detailed 2D and 3D drawings. Although the subscription is very expensive (approximately 262 EUR per month or 2,089 EUR per year), students and educators can use the program for free. Although it is not the most recommended option due to the high subscription cost for non-students/educators, if you still want to try it out and learn the basics, you can check the following links:

- **Tutorial for beginners**: In this tutorial series, *AutoCAD - Complete Tutorial for Beginners*, on YouTube, you can learn in detail the fundamentals of working with AutoCAD:
    - **Tutorial 1**: `https://www.youtube.com/watch?v=tHrfxjgFQt8`
    - **Tutorial 2**: `https://www.youtube.com/watch?v=c1kGuiYEHh0`
- **AutoCAD for laser cutting**: You can find a starter kit for working with AutoCAD at `https://www.ponoko.com/starter-kits/autocad#autocad_section_6`.

Inkscape

Inkscape is an open-source vector graphics software that can be used on Windows, macOS, and Linux. It allows you to create a wide variety of professional-quality graphics such as illustrations, icons, logos, diagrams, maps, and web graphics for free. To learn how to use Inkscape, you can access a wide variety of tutorials both in written and video formats on their website:

- **List of written and video tutorials**: For a list of written and video tutorials that can help you to get started with Inkscape, check the following link: `https://inkscape.org/learn/`.
- **Basic usage of Inkscape**: For an easy, short introduction on how to use Inkscape for designing, check the *Inkscape Lesson 1 - Interface and Basic Drawing* video on YouTube: `https://www.youtube.com/watch?v=8f0l1wdiW7g`.
- **Laser cutting with Inkscape:** Here is a well-written guideline on how to prepare a design for laser cutting on Inkscape: `https://www.sculpteo.com/en/tutorial/prepare-your-model-laser-cutting-inkscape/`.
- **Laser cutting with Inkscape video tutorial**: If you prefer video formats, however, the YouTube video, *Designing a Laser Cut Tabbed Box Using Inkscape*, provides an easy step-by-step tutorial for making a box for laser cutting in Inkscape: `https://www.youtube.com/watch?v=A1FI15Eq4PQ`.

Laser cutter features

Once you have gotten a good idea about the different programs and how to use them, you can design the case for your prototype. Now, let's discuss the different features of a laser cutter, one by one:

- **Cutting**: This refers to the action of cutting an outline and a shape from a piece of material.
- **Raster engraving**: In raster engraving, the laser sweeps back and forth across the working area, and individually engraves each pixel onto the base material of the workpiece.
- **Vector engraving**: In this type of engraving, the laser engraves the workpiece continuously and more smoothly. This engraving is smoother and much quicker than raster engraving and is generally used when working with flat surfaces.

For a better understanding of how raster engraving looks different than vector engraving, you can check the YouTube video, *Are vector and raster engraving different?* here: `https://www.youtube.com/watch?v=rZqUQ0w9vp4`.

Stroke color guide

To access the different features of the laser cutter (cutting, raster engraving, and vector engraving), you need to specify different outline colors for your shapes:

- **Red**: Cutting
- **Black**: Raster engraving
- **Blue**: Vector engraving

The colors may vary for different types of laser cutters. Laser cutting providers will specify which colors to use for their exact laser cutter model.

Finally, if you still find it difficult to design your case in any of those programs or just want to save some time, you can use this case generator to create the design: `https://en.makercase.com`.

Using 3D printing to build custom cases

In the previous sections of this chapter, you learned how to build a case for your prototype using household boxes and laser cutting technology. In this section, you will learn about a third option that you can use to produce nice-looking cases for your prototype, 3D printing:

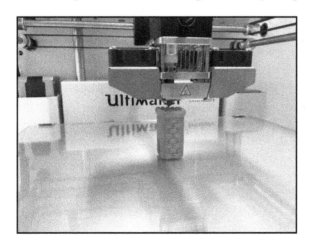

3D printer printing a tower

(source: https://pxhere.com/en/photo/605267)

3D printing is a technology that allows you to create a three-dimensional physical object from a digital file. It works using additive processes—that means processes in which an object is created by adding successive layers of materials (as opposed to removing materials) until the object is completed. The greatest advantage of using this technology is that it allows you to produce complex shapes using less material and more cost-effectively than when applying other manufacturing methods such a forging, molding, and sculpting. If you would like to use 3D printing to make cases for your prototypes, you can access this technology by either buying your own 3D printer, using 3D printers available at FabLabs, or using 3D printing services.

Buying a 3D printer – things to consider

Nowadays, there are more and more consumer-friendly 3D printers in the market and, fortunately, their price is within an affordable range. Printers targeted at consumers and hobbyists can be purchased for 250 EUR and more. The cheaper options may not offer the quality you are looking for, but you can still get a high-quality printer for about 1,000 EUR (which is still pricey but way more affordable than, for example, a laser cutter).

If you are considering buying a 3D printer, here a few things to take into account when choosing one:

- **Purpose of use**: When choosing which printer to buy, think about how you plan to use the 3D printer. If you are going to use the printer mostly for printing cases for your prototypes and you do not need the cases to be of outstanding quality, then you may not need the biggest and most expensive printer in the market. If you, however, plan to use the printer for more professional purposes and need the printer to, for example, print multiple objects at once, then you might want to choose a printer with a larger build area.
- **Size of the objects**: As mentioned briefly, it is necessary to take into account the size of the objects you plan to print to select a printer that suits your needs. If you plan to print larger objects, you need to make sure that the printer's build area (that is, the size, in three dimensions, of the largest object that can be printed) is big enough to be able to print them.

- **Materials used**: Most of the desktop 3D printers use a technique in which plastic filament is melted, extruded, and solidified to form the object. The two most common types of filament are **Acrylonitrile Butadiene Styrene (ABS)** and **Polylactic Acid (PLA)**. Following is a table to help you to better understand the advantages and disadvantages of each:

Material	Pros	Cons
ABS	• Low cost • High impact and wear resistance • High temperatures resistance • Less oozing and stringing • Smooth finish • Suitable for outdoors	• Needs a heated print bed • Produces fumes and odor, which makes it necessary to be in an open space or have very good ventilation • A tendency to shrink, which affects dimensional accuracy
PLA	• Low cost • Good strength • Good dimensional accuracy • Smooth look • Prints at low temperatures • No heated print bed necessary • Biodegradable	• Low heat resistance • May need cooling fans to prevent oozing • The filament can be brittle and break • Not suitable for outdoors (sunlight exposure)

Source: https://www.simplify3d.com/support/materials-guide/

Plastic filament comes in two diameters—1.85 mm and 3 mm—and is generally sold in spools of 1 kg at prices that range between 20–50 EUR. When you buy the filament for your printer, make sure that it is the right diameter for your printer and that the spool is the right size as well:

3D printing filament in different colors with models created using the filament
(https://en.wikipedia.org/wiki/3D_printing_filament#/media/File:3D_Printing_Materials_(16863368275).jpg) by Maurizio Pesce is licensed under CC-BY 2.0
(https://creativecommons.org/licenses/by/2.0/)

In addition to plastic, there are also a wide variety of other materials that can be used, such as metal (for example, bronze and copper), wood, nylon, and UV-luminescent filaments. With these materials, it is important to note that, since they have different melting points, to be able to use them, it will be necessary to buy a 3D printer that is either designed specifically for them or that allows users to control the extruder temperature:

3D printed metal parts (https://commons.wikimedia.org/wiki/File:3D_printed_metal_parts.jpg) by Loran Mak is licensed under CC BY-SA 4.0 (https://creativecommons.org/licenses/by-sa/4.0/deed.en)

In the following, we will have a look at two distinctive features of 3D printers: the resolution quality and the ability to print multi-colored objects:

- **Resolution quality**: In 3D printing, resolution equals layer height. The lower the number, the thinner the layer, and the thinner the layer, the finer the detail—which also means the higher the resolution. Most of the 3D printers in the market can print at a resolution of 200 microns (which produces decent-quality prints) and many can also print at 100 microns (which produces good-quality prints). If you want higher resolutions, please keep in mind that the price of the printer will likely go up and that more time will be required for printing.
- **Colors used**: Some 3D printers allow users to print objects in two or more colors but the most common are dual-extruder models. Although printing with different colors can be nice, the process for multicolored printing can be a little bit more complicated.

The previous criteria are meant to give you a rough idea of all of the things that you might want to further research if you decide on buying a 3D printer. There are many articles online with information and 3D printer reviews that can also provide you with a better idea of the features of different printers and their performance, and, in this way, help you to find one that suits your needs.

Using 3D printers in FabLabs

If buying a 3D printer is not an option for you, then FabLabs can help as they are usually equipped with 3D printers. If you intend to use the 3D printer for creating a case without any special features, then most FabLabs will have a printer that suits your needs. Just note that, as with the laser cutters, you will most likely need a membership and different types of membership will grant you access to different machines.

Exploring 3D printing services

Another option that you can check is 3D printing services. 3D printing services will print your design in the materials that you specify and send the printed object to you. As with laser cutting services, you just need to upload your 3D model, choose the desired material and place the order. They will then print the object and ship it to you. If you don't know how to make a 3D model for printing, some printing services also offer professional help or tutorials to guide you through the process:

3D printed model of stainless steel fastener prototypes by industrial equipment SLM 280 (https://www.flickr.com/photos/146658316@N02/40129461373) by Top3DShop is licensed under CC BY 2.0

A big advantage of using 3D printing services is that they often have equipment that allows printing in materials other than plastic. As you can imagine, 3D printers for special materials are more expensive, which might make it difficult to buy one and might not be available in FabLabs. 3D printing services allow you to access special printers for a fraction of the cost of buying one. There are many 3D printing services available but one of the most important and widely known 3D printing services is Shapeways.

Shapeways (`https://www.shapeways.com`) offers 3D printing in a wide variety of materials. You can order prints in plastic, TPU (thermoplastic polyurethane), metals (such as steel, aluminum, bronze, brass, gold, and silver) and even sandstone. The price for printing an object will greatly depend on the size and material but, just to give you an idea, minimum prices range from 5 EUR (the minimum price for printing in plastic) to 80 EUR (the minimum price for printing in platinum—but chances are that you are not thinking about printing a case for your prototype in this material).

Following is a table to give you a better idea of some of the materials available and what kind of objects those materials are often used for:

Material	Main properties	Frequently used for
ABS	Durable, strong, high temperatures resistance, and outdoors-friendly	Lego, phone cases, toys, sports equipment, car phone mounts, and toys
PLA	Odorless, biodegradable, less energy necessary, low heat resistance, and sometimes brittle	Prototype parts, models, and biodegradable medical implants
Nylon	Strong, lightweight, durable, flexible, and wear-resistant	Cases, mechanical parts, tools, prosthetics, fixtures, tech accessories, prototypes, figurines, and jewelry
TPU	Flexible and rubber-like, durable, abrasion-resistant, and smooth feeding properties	Automotive, medical, robotics, footwear, helmet interiors, and toys
Wood	Versatile, durable, and contains real wood fibers and has a wood scent	Boxes, figurines, cases, tables, and chairs
Sandstone	Very brittle, low strength, and great for multicolor printing	Figurines, architecture models, and medical models
Steel	Affordable, good strength, and many colors available	Tools, accessories, home decor, and small sculptures; not suited for industrial load-bearing applications

Aluminum	Light but strong, allows high accuracy, good corrosion resistance, outdoor-friendly, and high electrical and thermal conductivity	Mechanical parts, tools, fixtures, bike and drone accessories, and structural components
Brass	Showcases intricate details and has a professional and finished look and feel	Jewelry, fashion accessories, home decor, prototypes, and miniatures
Bronze	Subtle marbling effect, vintage look, showcases intricate details, and has a finished look and feel	Jewelry, fashion accessories, home decor, prototypes, and miniatures

Source: https://www.shapeways.com/materials

Now that you have a better idea of what 3D printing is, how you can have access to a 3D printer and what kind of materials you could use for your prototype case, it is time to discuss how you can design your case's 3D model for printing.

Building a case for 3D printing in Tinkercad

Using 3D printing to build a case for your prototype is a great choice. It not only allows choosing between various colors and levels of shininess and flexibility but, most important of all, gives you tight control over the form. As shown in previous sections, there are many options on how to get your hands on a 3D printer. You can either buy a 3D printer yourself, use a maker space (FabLab) for a monthly fee, or use an online 3D printing service such as Shapeways (https://www.shapeways.com) for a one-time cost.

In the following section, I want to give you a tool that makes it incredibly easy to design your own cases for your prototypes.

Tinkercad (https://tinkercad.com) is a web-based tool developed by US company, Autodesk, a company at the forefront of 3D design tools and that developed Fusion 360 (https://www.autodesk.com/products/fusion-360/overview), a professional tool suitable to be used for more advanced 3D modeling tasks.

Tinkercad is targeted at makers and only ships with the features needed to create 3D models based on geometric primitives (like boxes and spheres) in your browser and export them for 3D printing. Having tried many other 3D applications (most of them targeting industrial applications) before, I was delighted to find a tool that does exactly what I need, nothing more and nothing less.

Being able to create your own prototype cases can bring the perceived level to a whole new level. While we probably all agree that what is inside is the most important part (true for humans as well as electronic prototypes), having a nice shell around your prototype makes a huge difference though.

Creating variations of other people's work in Tinkercad is possible using **remixes**. Remixing another user's project allows you to tweak each and every setting while still giving attribution to the other user. Being able to tweak the project of other users is incredibly useful. Every 3D printer prints a little bit different and often 3D models must be slightly adapted.

For people who do not have access to a 3D printer, another feature might come in handy: direct ordering via the Tinkercad website. Once you are happy with your prototype case, you can order it without leaving the site. I have not used this feature, yet, but I am sure it is worth a try.

Starting with Tinkercad is easy, not only for people who have used other 3D software before but also for complete 3D beginners. There are a few concepts which are important for you to create your prototype cases in Tinkercad:

- **Primitives**: Most prototype cases can be built using the basic geometric shapes, such as cube and cylinder.
- **Transformations**: Objects can be resized or moved in the x, y, or z direction.
- **Additive behavior**: When you export your case, all objects that overlap or touch each other will be merged to become one bigger object.
- **Holes**: Each shape can be set up as a hole. You would, for example, use a hole object to create an opening for your cables or user interface elements such as buttons and knobs.
- **The grid**: Tinkercad features a grid, which will result in a snap-in behavior when you resize or move objects around. Set it up in a way that makes sense for you. Often, 1 mm is good, but sometimes you will need more control, so choosing a 0.1 mm grid might be better.

Let's build a case together. Please follow along with my explanations. I will not be covering the steps needed to set up an account on Tinkercad here. Also, the following instructions should not be seen as step-by-step instructions but should give you a general idea of what the process looks like. Simply reading the instructions might already give you an idea if you want to try using Tinkercad yourself.

The first step needed in every 3D design is a sketch with pen and paper. Maybe you prefer a different way of sketching, but for me, using pen and paper is the fastest. Putting your basic idea on paper will make sure that you thought your idea through just enough. We are not planning to build a house here, where just enough would definitely not be enough. We just want to make sure that we don't lose time creating multiple iterations of the 3D design because we made a mistake that could easily have been prevented by sketching it out with pen and paper first.

Just like in our paper-case for the smart productivity cube, we want an outer cube with an opening at the back for our USB cable. The biggest difference needed compared to the paper cube is the opening mechanism. The material used by the 3D printer is not foldable, so we have to come up with a different mechanism—a lid. The lid must fulfill the following functions:

- It should not slip inside the cube.
- It should sit tight, but not too tight.
- The opening for the USB cable must be big enough for the cable to fit through.

Additionally, we need to make sure that our breadboard-construction fit into the cube. Optimally, this would be replaced by a soldered circuit board (to save space and increase stability), but to keep it simple, let's just continue to use the breadboard-construction.

 If you plan on using any of the 3D print-on-demand services, make sure to double-check the measurements of your model correctly. Nothing is more frustrating than receiving your 3D model with wrong dimensions.

The first thing we need to do to create a case for the smart productivity cube we built in Chapter 7, *Building a Smart Productivity Cube,* is to add a cube to the scene:

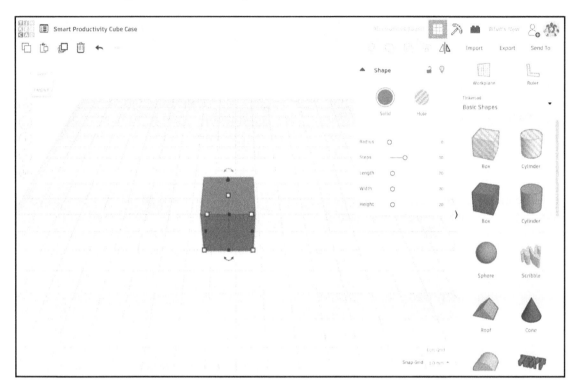

The cube was dragged from the sidebar to the scene

The cube must then be scaled to have the correct dimensions:

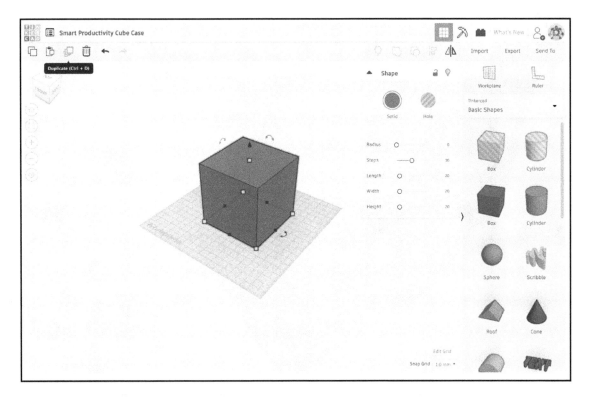

Scaled cube

When dragging one of the handles to scale the cube, a tooltip shows the current dimensions:

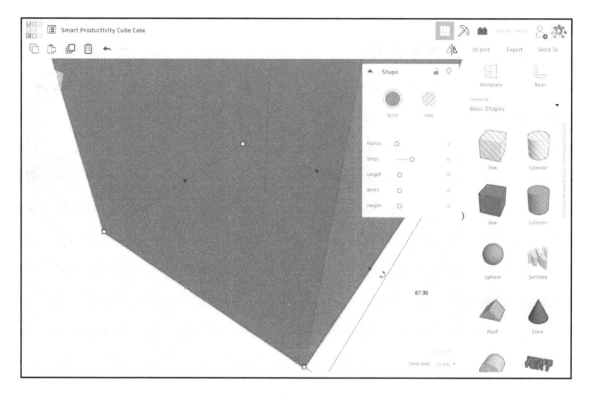

Size tooltip

Now, we need to do the following:

1. **Clone the cube**: Click the clone icon in the menu bar (third icon from left).
2. **Move the cloned cube**: Moving it makes it is easier to see the difference to the other cube.
3. **Resize the cloned cube**: It should be one 1 mm smaller on each side:

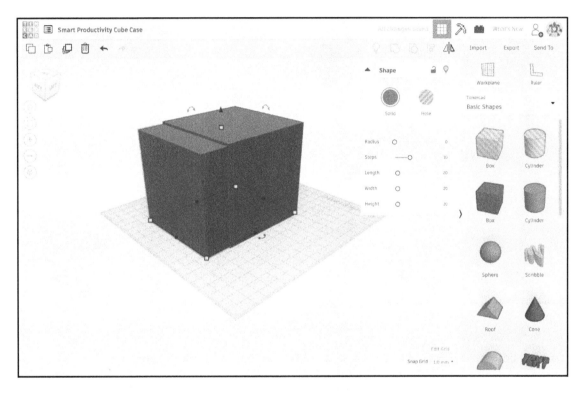

The cloned (smaller) cube

Now, select the smaller cube and press **Hole** in the shape options on the right to convert the object into a hole object. It will then be subtracted from the other object:

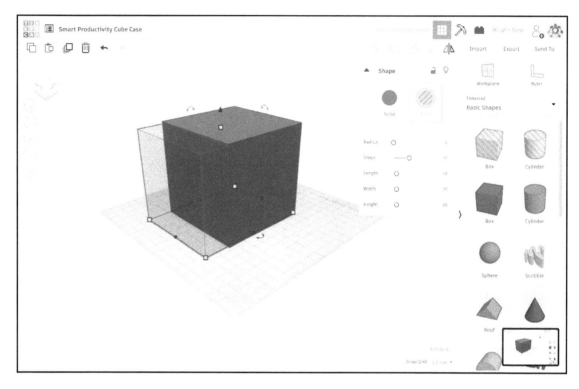

Using the hole object creating a case for our prototype is easy as pie

To get a better overview of our cube with a hole, we can now create a group:

1. **Create selection**: Select both cubes with your mouse or press *cmd + A/Ctrl + A*
2. **Create group**: Click the **Group** icon in the menu bar or press *cmd + G/Ctrl + G* to group both cubes together. If you have ever used Photoshop or a similar tool, you are probably used to the concept of groups.

Now, you will see the grouped cube. It looks more realistic and makes it easier to model the missing parts:

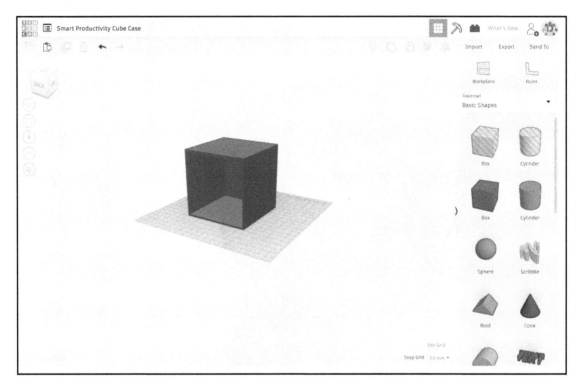

Both cubes grouped together

 If, later on, you want to make changes to an object in a group, simply mark the grouped object and press **Ungroup**. You can also double-click the object to access the objects inside the group.

We then have to create the lid by applying the same technique we used for the body of the box as the inner part, combined with a solid box as the outer part:

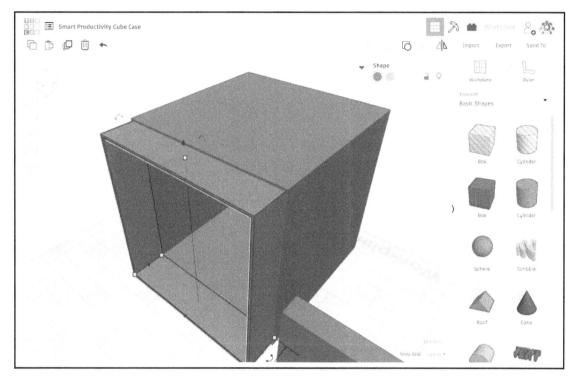

Multiple cubes will be combined to form the lid

After combining the two lid-parts, we can rotate and move the lid in place to see whether it fits:

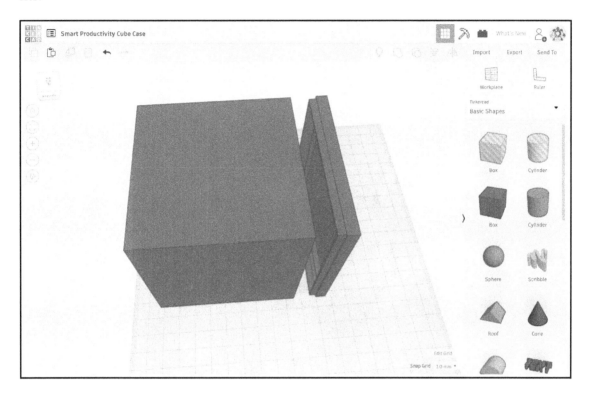

We are nearly done. We can place our prototype inside the cube now and open and close the lid:

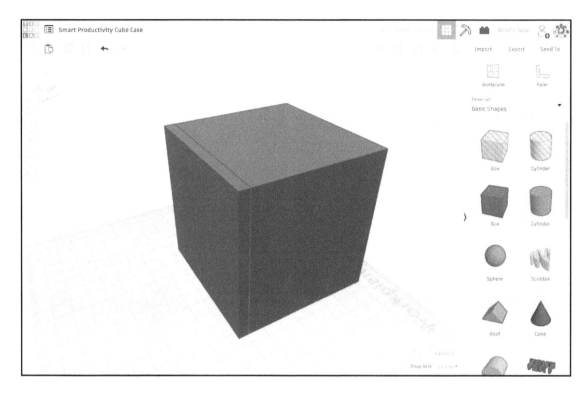

Closed cube

 When working on the computer it is easy to forget, that 3D printing is a physical process and models might be a little bit bigger or smaller when being printed. You should plan in a little tolerance, so that cube body and lid fit together.

We still have to add one thing to our prototype case: an opening for the USB cable. We just need to do a couple of steps:

1. Add another hole object
2. Scale it so the smaller side of a USB cable can fit through
3. Move it inside the lid:

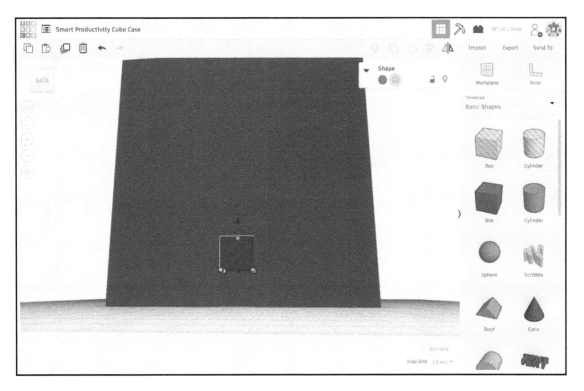

Hole object positioned in the lid

The finished cube with a hole for the USB cable looks like this then:

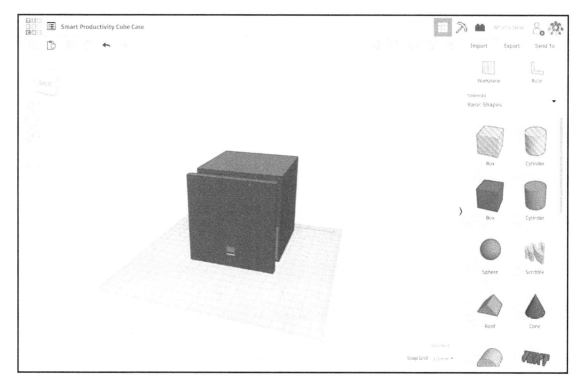

Cube body and lid with hole for the USB cable

Before exporting the print, there is one thing left to do, which will make it easier to print the parts—rotate the parts so that the filled side lies on the ground:

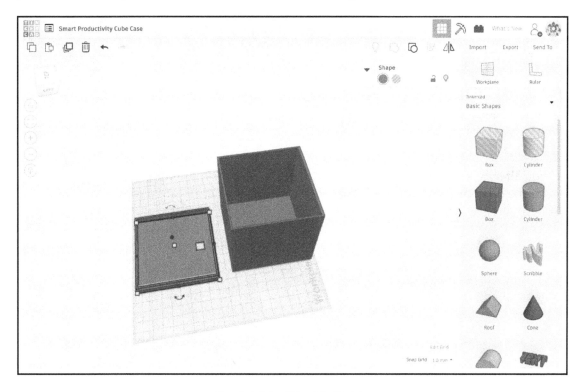

Rotated cube body and lid, ready for 3D printing

Great! That's it. You could now press on **Export** to either save the model to disk in STL format or order it directly via one of the partner shops, in any material or color that you like.

If you wanted to make your prototype even more stable, you could also create a holder for your electronics where everything fits nice and tidy into the cube body.

Evolution of a circuit – from breadboard to PCB

In the previous sections, different ways to build cases for your prototype were discussed to help you give your prototype a polished look. In this section, the focus is shifted to what is inside the case and a guideline on how to go from using a breadboard to making a professional **Printed Circuit Board** (**PCB**) is introduced.

Breadboards

As you already know, a breadboard is a simple construction base for prototyping electronics. Although breadboards come in many different sizes and designs, in general, they are composed of bus strips (the power rails marked with a blue and red line used to connect a power source and ground for the electronic components) and terminal strips (the main area used to hold most of the electronic components):

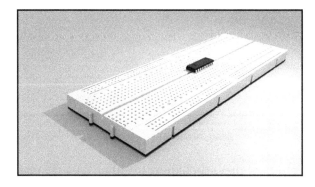

Breadboard (final render pic on breadboard (https://commons.wikimedia.org/wiki/File:Final_render_pic_on_breadboard.png) by Cz-David is licensed under CC BY-SA 3.0 (https://creativecommons.org/licenses/by-sa/3.0/)

Because breadboards are composed of metal pieces designed to grab onto the legs of any components pushed through the breadboard holes, they do not require soldering. This makes it easy to create circuits very fast and makes it possible to reuse the breadboard. Therefore, breadboards are quite useful when you are creating temporary prototypes and experimenting with circuit design. However, precisely because of their solderless quality, they are not very stable and it is not uncommon to have a cable get ripped out. For this reason, when you are past the stage of experimenting and have built a circuit that should be kept, it is highly recommended to use a solderable option to secure the components and cables.

Solderable breadboards

Once you have built a circuit that you would like to keep, it is time to move to a solderable breadboard. There are many solderable breadboards available in the market, but a recommendable option is a solderable breadboard by Adafruit:

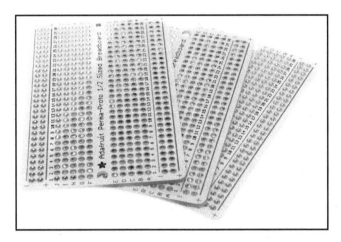

Adafruit Perma-Proto half-size breadboard PCB (https://www.flickr.com/photos/33504192@N00/14151911408) by oomlout is licensed under CC BY-SA 2.0 (https://creativecommons.org/licenses/by-sa/2.0/)

The solderable breadboard by Adafruit looks pretty much like a solderless breadboard. The top side has the same markings for the bus strips and the numbers for terminal strips of a classic breadboard. This makes it very easy for you to transfer the circuit created in the breadboard to this one. The diameter of the holes allows you to fit big leads and two mounting holes make it easy to attach it to your case. For this reason, it is one of the easiest to use and preferred options by many people.

The solderable breadboard is useful to make your circuit design permanent. However, it still requires that you manually place the circuit and solder it. Therefore, it is not the best option if you want to replicate your circuit multiple times. Instead, at this stage in the iteration process, it is recommended to use printed circuit boards.

Printed circuit boards

A printed circuit board is a board that connects electronic components. It is composed of conductive tracks, pads, and other features etched from copper sheets and laminated on a non-conductive substrate. The different components of the board are usually soldered to ensure that they are secured (`https://www.pcbtrain.co.uk/blog/what-is-a-printed-circuit-board`).

There are many advantages of using PCBs but some of the main benefits are as follows:

- **Time-saving**: Printed circuit boards make it easy to assemble circuits in less time. The PCB typically contains all of the traces you defined in the design using circuit design software, and, therefore, you do not need to solder the cables anymore (the cables are already part of the PCB). You only need to solder the electronic components.
- **Easily repairable**: PCBs have labels on the board indicating components and polarities. This makes it easy to check and repair them.
- **Space-saving**: As the PCB uses copper tracks to connect components instead of wires, it allows you to have a large number of electronic components without making the circuit bulky. Also, as electronic components can be arranged in a compressed and efficient way, the PCB enables you to make big and complicated electronic circuits without taking a lot of space.
- **Connects small-size components**: The PCB makes it possible to connect very small components that would otherwise be impossible to connect with wires.
- **Affordable**: Ordering great quantities of a PCBs involves a relatively low cost (source: `https://www.edgefx.in/advantages-using-printed-circuit-board-pcb/`).

Due to the previous advantages, PCBs are a good option if you want to start producing your circuit in greater quantities. However, they require additional design effort to lay out the circuit. In the next section, some of the most known and widely used **Electronics Design Automation** (**EDA**) software will be introduced for you to get a better idea about how to make the circuit board design.

Electronics design automation or electronic computer-aided design software

EDA or **Electronic Computer-Aided Design** (**ECAD**) software are tools for designing electronic systems such as printed circuit boards and integrated circuits. You can find the following short explanation of some of the most widely known and used ECAD tools.

Eagle

Easily Applicable Graphical Layout Editor (**Eagle**) (`https://www.autodesk.com/products/eagle/overview`) is one of the oldest and most popular PCB design tools. It has a simple interface and provides a library with a great number of components. It enables both beginner and advanced designs, it is not so difficult to learn, and it has a large online community with tutorials that can help you to get started. All of this makes it suitable for both hobbyists, makers, and more professional users.

Eagle offers two subscription types: standard and premium. Access to the standard version costs about 18 EUR per month or 131 EUR per year. The premium subscription offers more features but, as a result, is much pricier (approximately 65 EUR per month and 520 EUR per year). However, Eagle also provides access to a limited version of the software for hobbyists and makers for free. This version includes 2 schematic sheets, 2 signal layers, and an 80 cm^2 board area, which is enough for most hobbyists and makers.

In case you would like to check this software out and get started with it, here is a basic tutorial: *Getting started with EAGLE* on YouTube at `https://www.youtube.com/watch?v=SkxbnIypGwY`.

Altium designer

Altium (`https://www.altium.com/altium-designer/`) is one of the most popular high-end PCB design applications today. Its focus is to enable users to produce complex final board designs for fabrication rapidly. Therefore, it is very comprehensive but also not so easy to use. Some of the features that differentiate it from other options include its capacity to produce high-end product designs, team-based design development, good graphics processing power, and a 3D visual engine that makes it easier to figure out circuitry problems. Price-wise, Altium is on the expensive side, which has limited its use in the hobbyist and DIY community. Overall, although it is a powerful tool, it is not really recommended unless you will use it in a professional setting in which you have to design very complex circuits and work with large teams on PCB designs.

If you have not used Altium before, here is a video that will help you to learn how to design PCBs in this software:

- **Altium Tutorial Part 1 - Schematics**: `https://www.youtube.com/watch?v=XqE-AoOLRJE`
- **Altium Tutorial Part 2 - PCB**: `https://www.youtube.com/watch?v=gg5Zj650JCo`

KiCAD

KiCAD (`http://www.kicad-pcb.org/`) is an open source, free EDA software that offers many high-end features similar to those in professional PCB design packages. It is not very difficult to use and has its own library with most electrical components, which makes it a good option for beginners. One downside is that there are different modules, which are not so easy to navigate and take some time to get used to.

KiCAD can run in different operating systems, including Windows, macOS, and Linux and is available in multiple languages. If you want to check out KiCAD, you should have a look at the following playlist:

- **An Intro to KiCad**: `https://bit.ly/2FWmKwU`

Fritzing

Fritzing is an open source tool that is mainly used by hobbyists to document their breadboard designs. The circuit graphics in the hands-on part of this book were made in Fritzing as well. Fritzing offers a service called Fritzing Fab, which makes it possible to order a PCB from a Fritzing design.

Learning how to use Fritzing is very straightforward, but, due to its simplicity, it also has its limitations. While the development of Fritzing has been on hold for a few years, some new features and bug-fixes can be expected in 2019 (see `https://github.com/fritzing/fritzing-app/issues/3443`). It definitely is the easiest option to go from breadboard design to PCB, but as mentioned before, due to its simplicity, it has its limitations and quirks.

To get an idea of how to create a PCB design in Fritzing, you can watch the video, *One minute Arduino Shield design*, on YouTube: `https://www.youtube.com/watch?v=eHU-pF5gSnQ`.

Ordering a PCB

All of the previously listed services make it possible to design a PCB which can then be ordered by a PCB service. They all work similarly:

1. Upload your PCB design.
2. Fine-tune the parameters for your order (for example, the color of the PCB).
3. Order the PCB.

There is a multitude of PCB production services to choose from. Some of the options are as follows:

- **Euro circuits** (https://www.eurocircuits.de)
- **EasyEDA** (https://easyeda.com/order)
- **Seeed studio PCB** (https://www.seeedstudio.io/fusion_pcb.html)
- **Fritzing fab** (https://aisler.net/partners/fritzing), if your design was made in Fritzing

What you will receive when placing an order in any of those services is a PCB with routes and holes to place your components in. You will still need to hand solder the components to the PCB, which is fine for a couple of prototypes, but definitely not tolerable if you are thinking about (small) batch production, as it simply consumes too much time.

Pick and place

The next and final step is to also automate the process of placing the electronic components onto the PCB as well as soldering. For the placement, so-called pick and place machines are used. They have a robotic arm, which is programmed to pick up the tiny components, place them in the right holes of the pre-manufactured PCB, and, once all of the components are on it, it is soldered (not by hand, but often with the reflow technique):

- **Ultra High-Speed Pick and Place Machine**: https://www.youtube.com/watch?v=8G7pUqRZjU8
- **What Is Reflow Soldering?**: https://www.youtube.com/watch?v=eOUf59iut3s

Summary

In this chapter, you were introduced to different techniques and technologies to create cases for your prototypes. We had a look at laser cutting as well as 3D printing. Using the web-based tool Tinkercad, we re-built the case for our smart productivity cube from scratch and prepared it for 3D printing.

Lastly, you were introduced to the most common software used for PCB design, where to order PCBs, and you were given an overview on what other machines might be needed to produce bigger batches of hardware using pick and place machines. Even if producing prototypes in series seems far ahead, you now have a better idea of what the next steps could involve if you want to professionalize your prototypes.

I would like to thank you for following me on this journey on getting to know physical prototyping with Arduino and MQTT. I hope I was able to communicate the basic concepts of MQTT clearly and that you feel comfortable using it in your next projects.

But most of all, I hope you feel inspired; inspired by the power of creating functional prototypes with very little code, just by combining existing examples, a few electronic components, and household items; inspired by adding internet-connectivity using MQTT to your prototypes. There are a gazillion possibilities that you can create with it and, because of MQTT's open nature, it will open up new ways to interact with your prototypes using third-party services, libraries, and tools.

What will you make next?

I would be happy if you leave your feedback (positive or negative) as an issue on the GitHub page of this book (`https://github.com/PacktPublishing/Hands-on-Internet-of-Things-with-MQTT`). If you have any questions about follow-up projects, I am happy to help.

Assessments

Chapter 1: The Internet of Things in a Nutshell

1. Some technologies associated with IoT are smart homes, smart cars, and IIoT/Industry 4.0.
2. Voice user interfaces such as Alexa and Siri are used more and more to control internet-connected devices.
3. It does not make sense to make every device smart. The more features a device has, the more likely it is for problems to appear. Would you rather have a traditional vacuum cleaner that works 99% of the time, or a vacuum cleaner that only works when a wireless network connection is available but that shows its status in an app?
4. Prototypes are used to find out if your idea works or not. They can be developed in a quick and dirty way and are all about getting the desired result fast.
5. No! The beautiful thing about prototyping is that you don't need to be an expert in every area—especially in the Arduino community, where you will find code snippets and diagrams for most common sensors and actuators. Combining example snippets and adding a little bit of logic to the code might do the job for a first version.

Chapter 2: Basic Architecture of an IoT Prototype

1. We are living in exciting times! There are a lot of development boards available that are suitable for IoT prototyping. These include the Arduino MKR WiFi 1010, the Raspberry Pi 3 Model B+, M5Stack, NodeMCU, and the Particle Argon, Boron, and Xenon.

2. Depending on what you plan on building, picking a development board that runs on 3.3V or 5V might be better. If you want to work with Neopixel LEDs, for example, then picking a 5V development board is easier. But the trend is moving toward 3.3V. With a level shifter/logic level converter, you can use 5V modules together with a 3.3V development board, as well.

3. ZigBee, Thread, CoAP, and MQTT are used for IoT communication.

4. Some development boards support over-the-air updates, which means that their firmware can be updated without a cable connection, using a wireless internet connection instead.

Chapter 3: Getting Started with MQTT

1. You can send MQTT messages via Mosquitto using the `mosquitto_pub -t "/test" -m "Your message"` command.

2. You can subscribe to MQTT messages via Mosquitto using the `mosquitto_sub -t "/test"` command.

3. The `-t` parameter stands for `topic`.

4. The `-m` parameter, which you have to use when sending messages, stands for `message`.

5. Please explore for yourself.

Chapter 4: Setting Up a Lab Environment

1. To publish an MQTT message with a `Hello` payload to the `/test` topic, you have to run `mosquitto_pub -t "/test" -m "Hello!"` in the Terminal.

2. To subscribe to MQTT messages that are sent to the `/test` channel, you have to run `mosquitto_sub -t "/test"` in the Terminal.

3. The `-t` command-line flag stands for `topic`.

4. The `-m` command-line flag stands for `message`.

5. This is an exploratory exercise. Go to `https://shiftr.io/explore` and have a look around.

6. This is an exploratory exercise.

7. This is an exploratory exercise. If you are having problems, note that, in the hands-on chapters, we will make use of third-party apps to send and receive MQTT messages together.

Chapter 5: Building Your Own Automatic Pet Food Dispenser

1. The # character is called a multi-level wildcard. By using it, we subscribe to all subtopics of the topic.
2. In this project, we are using the public namespace with the `try`/`try` login credentials. It is neither secure nor private, and we just use it because it is easier and more tolerable for a prototype. If you want more security and privacy for your data, you need to use a private channel on Shiftr (or any other MQTT server with private channel functionality).
3. If Shiftr stops working at some point, you can just move on to another (free) MQTT server (a list can be found at `https://github.com/mqtt/mqtt.github.io/wiki/public_brokers`). This is the beauty of MQTT: there are many implementations, and you can easily move on to another provider or simply create your own (local) server.
4. You can control the smart pet food dispenser from any MQTT client. MQTT clients come in many shapes and sizes. You can control it using the Terminal (for example, using Mosquitto, which we installed in `Chapter 3`, *Getting Started with MQTT*), using iOS or Android apps, or using macOS and Windows apps. You can also create a new Arduino project that also uses MQTT, and easily let the two Arduinos communicate using MQTT in the same way.
5. Writing as little code as possible is the most efficient way of making your ideas a reality. When you find tiny examples or code snippets that solve parts of your problems, you should try to combine them instead of writing all of the code from scratch. You will be faster this way: you don't need to be an expert, and mostly using code that has been used by other people means that it will probably work when you use it.
6. You really don't need to be an expert to build breathtaking prototypes. In the physical prototyping workshops I've held, many people who have never written code before built awesome prototypes anyway by combining examples, adding some code to glue them together, and making it look nice by spending time on building a case and hiding the (often ugly-looking) inner workings of the prototype.

7. Please don't bet the life of your pet on your prototype working. As stated before, it is not secure, can probably fail in other ways, and should just be used as a prototype.

8. There are many more things that you can build using just a servomotor and the techniques we used. Have a brainstorming session and try to come up with five ideas.

Chapter 6: Building a Smart E-Ink To-Do List

1. QoS stands for quality of service. It was introduced in `Chapter 3`, *Getting Started with MQTT*.

2. QoS 1 is a good compromise for our use case between reliability and performance. With QoS 1, we can be sure messages are delivered (if the other end also uses QoS 1 or 2).

3. MQTT is an open protocol, so there are libraries for every programming language, more or less. There are also a lot of pre-made third-party applications for Android, iOS, macOS, Windows, Linux, and the web, which you can use. All you have to do is specify the login credentials of the MQTT server you are using (in our case, `https://shiftr.io/`).

4. A client ID is basically the name of the MQTT client, as seen in the network. The MQTT server stores messages to be sent, as well as subscriptions for each client ID. Don't confuse this with the MQTT username and password: these are just for authentication.

5. There can be only one `loop` and one `setup` function for each Arduino sketch. If you combine various examples into one, you need to make sure that you integrate the code accordingly. Having two functions with the same name will result in a compile-time error.

6. There are many examples given in `Chapter 3`, *Getting Started with MQTT*, for MQTT apps on various platforms. You are missing out on all of the fun if you do not try them out. Third-party MQTT apps allow you to build your own user interfaces to control your smart devices without having to learn another programming language.

7. When using your own login on `https://shiftr.io/`, you can activate private namespaces, so other MQTT clients cannot see what you send and cannot interfere with your namespace.

Chapter 7: Build a Smart Productivity Cube, Part 1

1. Tilt switches contain a small metal ball inside that closes an electronic connection when held in one specific position. If the tilt switch is placed flat, the small metal ball can go either way; when you read it, the result might be either `HIGH` or `LOW`—you get random readings. When working with tilt switches, you should make sure that the small metal ball inside is always in a stable position by using appropriate angles in your construction.

Chapter 8: Building a Smart Productivity Cube, Part 2

1. MQTT apps for iOS and Android can be used to both display and send MQTT messages. Their purpose is to build a personal dashboard that allows you to consume all of the important information published by your MQTT devices and to control them by publishing messages as well.

2. To send just the activity time (without the activity name), you have to change the `client.publish("/tims-channel/cube/activity", text);` line to `client.publish("/tims-channel/cube/activity", timeSpend);`.

3. A good client ID, to be compatible with most MQTT clients and servers, should be no longer than 23 characters and only contain regular characters (a-z, 0-9, and -). Also, you should make sure that nobody else is using the same name on the same MQTT server. The easiest way is just by appending a few random digits to your client ID (for example, `your-name-184638`).

Other Books You May Enjoy

If you enjoyed this book, you may be interested in these other books by Packt:

Mastering Internet of Things
Peter Waher

ISBN: 978-1-78839-748-3

- Create your own project, run and debug it
- Master different communication patterns using the MQTT, HTTP, CoAP, LWM2M and XMPP protocols
- Build trust-based as hoc networks for open, secure and interoperable communication
- Explore the IoT Service Platform
- Manage the entire product life cycle of devices
- Understand and set up the security and privacy features required for your system
- Master interoperability, and how it is solved in the realms of HTTP, CoAP, LWM2M and XMPP

Internet of Things Programming Projects
Colin Dow

ISBN: 9-781-78913-480-3

- Install and set up a Raspberry Pi for IoT development
- Learn how to use a servo motor as an analog needle meter to read data
- Build a home security dashboard using an infrared motion detector
- Communicate with a web service that sends you a message when the doorbell rings
- Receive data and display it with an actuator connected to the Raspberry Pi
- Build an IoT robot car that is controlled through the internet

Leave a review - let other readers know what you think

Please share your thoughts on this book with others by leaving a review on the site that you bought it from. If you purchased the book from Amazon, please leave us an honest review on this book's Amazon page. This is vital so that other potential readers can see and use your unbiased opinion to make purchasing decisions, we can understand what our customers think about our products, and our authors can see your feedback on the title that they have worked with Packt to create. It will only take a few minutes of your time, but is valuable to other potential customers, our authors, and Packt. Thank you!

Index

building, laser cutting used 284, 285

D

deforestation, with TensorFlow
  reference link 21
design, generating for laser cutter
  about 288
  Adobe Illustrator 289
  AutoCAD 289
  CorelDRAW 288
  Inkscape 290
Device Cloud
  reference link 34
dynamic IP addresses 63

E

e-paper device
  accessing, via serial communication 190, 191,
    193, 195, 196
  accessing, with MQTT 198, 202, 203, 204, 206
  case, building 213, 215
  enhancements 213, 215
e-paper example
  modifying 188, 189
  simplifying 178, 181, 183, 184, 187
e-paper module
  connecting, to Arduino MKR WiFi 1010 171,
    173, 176, 177
  example, executing 171, 173, 176, 177
Easily Applicable Graphical Layout Editor (Eagle)
  about 315
  reference link 315
  subscription types 315
Electron
  URL 36
Electronic Computer-Aided Design (ECAD) 315
Electronic Design Automation (EDA) 314, 315

F

fabrication laboratory (fab lab)
  about 285
  accessing through 286, 287
  reference link 287
features, laser cutter
  accessing, with stroke color guide 291

cutting 290
  raster engraving 290
  vector engraving 290
Figma
  reference link 283
Formulor
  reference link 287
Fritzing
  about 316
  limitations 316
  reference link 316
front-plate interface
  basic shapes 283
  designing 281, 282, 284
  text rendering 283
  working, in cm/mm/in 283

G

Google Cloud
  URL 82
Google Flutter
  URL 36
Google
  URL 38
Graphical User Interface (GUI) 36
Ground (GND) 171, 228
Grove System
  reference link 56
GxEPD2 library 169

H

household boxes
  front-plate interface, designing 281, 282, 283,
    284
  using, as cases 279, 280, 281

I

If This Then That (IFTTT)
  URL 34
industrial Internet of Things
  exploring 18, 19, 20
industry 4.0
  exploring 18, 19, 20
Inkscape 290
Inkscape, tutorials

N

www.ingramcontent.com/pod-product-compliance
Lightning Source LLC
Chambersburg PA
CBHW080618060326
40690CB00021B/4733